Electronics Servicing

J.S. Anderson

To my favourite (and only) daughter, Sam

Butterworth-Heinemann
Linacre House, Jordan Hill, Oxford OX2 8DP
A division of Reed Educational and Professional Publishing Ltd

A member of the Reed Elsevier plc group

OXFORD BOSTON JOHANNESBURG
MELBOURNE NEW DELHI SINGAPORE

First published 1997

British Library Cataloguing in Publication Data
Anderson, J. S. (John Stuart), 1941–
Electronics servicing
1. Electronic apparatus and appliances – Maintenance and repair 2. Electronic circuits – Testing
I. Title
621.3′81548

ISBN 0 7506 3554 1

Typeset by Laser Words, Madras, India
Printed and bound in Great Britain by Bath Press, Avon

Contents

15 In Figure 1.13, if the battery voltage is 10 V and the current flowing is 250 mA the value of the resistance is

A 2.5 Ω
B 4 Ω
C 25 Ω
D 40 Ω

16 In Figure 1.13, if the resistance is 45 Ω and the current flowing is 0.2 A, the battery voltage, V is

A 2.25 V
B 9 V
C 90 V
D 225 V

17 In Figure 1.13, if the resistance is 25 Ω and the battery voltage is 10 V, the current flowing is

A 40 mA
B 400 mA
C 2.5 mA
D 250 mA

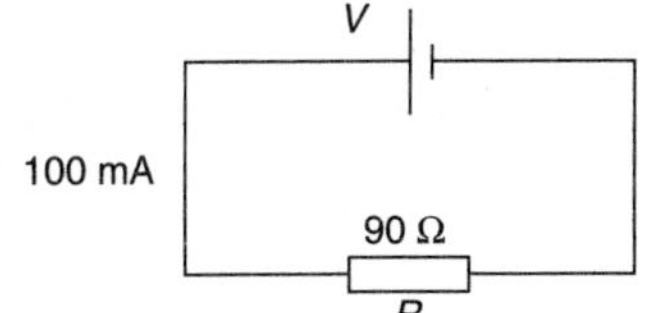

Figure 1.14

18 Refer to Figure 1.14, The battery voltage is now halved and the resistance doubled. The new current flowing is

A 25 mA
B 50 mA
C 250 mA
D 500 mA

19 In Figure 1.14, the battery voltage is now doubled and the resistance is halved. The new current flowing is

A 50 mA
B 100 mA
C 200 mA
D 400 mA

20 In Figure 1.14, what value of resistance would be required in order to reduce the current to 1 μA?

A 900 Ω
B 9 kΩ
C 900 kΩ
D 9 MΩ

21 A potential difference of 1 volt is equal to

A one amp per ohm
B one amp per second

C one amp-ohm
D one amp per joule

22 Power is measured in

A watts
B joules
C coulombs
D units

23 The power rating of a heating element of resistance 5.76 Ω connected to the 240 V mains is

A 41.7 W
B 1382.4 W
C 1 kW
D 10 kW

24 The power developed in a circuit using energy at the rate of 500 joules per second is

A 5 W
B 50 W
C 500 W
D 5 kW

25 The maximum current that could safely be carried through a 5 W, 18 Ω resistor is approximately

A 0.53 A
B 1.9 A
C 3.6 A
D 90 A

2 Serial and parallel resistor networks

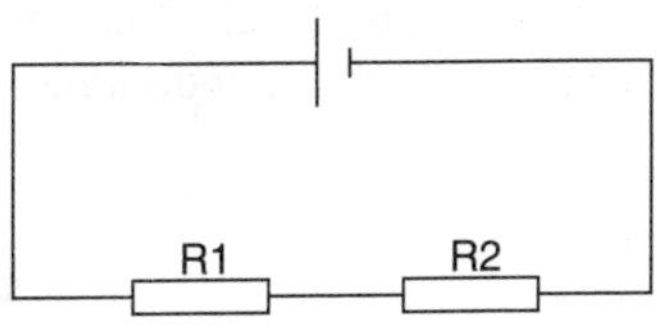

Figure 2.1 Resistors in series

In the last chapter it was shown how Ohm's law can be used to calculate one of the three values of I, V and R, provided the other two are known. The simple circuit given included just one resistor, but what if there's two or more? If the resistors are placed one after the other as shown in Figure 2.1 this is termed 'resistors in series', and all you have to do, in order to get the total effective resistance, is add them together.

Say R1 = 5 Ω and R2 = 7 Ω and the supply is a 12 V car battery, to work out the current, first find the total resistance, i.e.

$$R_{tot} = R_1 + R_2$$
$$= 5 + 7 = 12\,\Omega$$

From Ohm's law, $I = V/R = 12/12 = 1\,\text{A}$

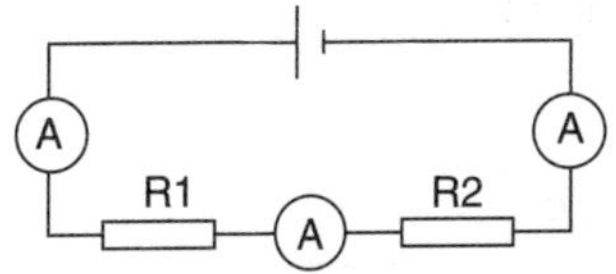

Figure 2.2 In this series circuit, each ammeter reads the same value

If there are three resistors in series, they are also simply added together in order to obtain the total resistance, R_{tot}, i.e.

$$R_{tot} = \text{R1} + \text{R2} + \text{R3} + \ldots$$

One important point to remember is that the current in a series circuit does not get 'used up'. An electric current is similar to water flowing through a pipe, what goes in at one end, must come out of the other. Hence in Figure 2.2, all the ammeters read the same value.

Summary of important facts about resistors in series

1. The current is the same through all resistors.
2. The total p.d. = the sum of the individual p.d.s.
3. Individual p.d.s are directly proportional to individual resistance.
4. The total resistance is greater than the greatest individual resistance.
5. The total resistance = sum of individual resistances.

Resistors in parallel

Figure 2.3 shows a parallel arrangement of two resistors. It should be obvious that the voltage is the same across each resistor since they are both connected together.

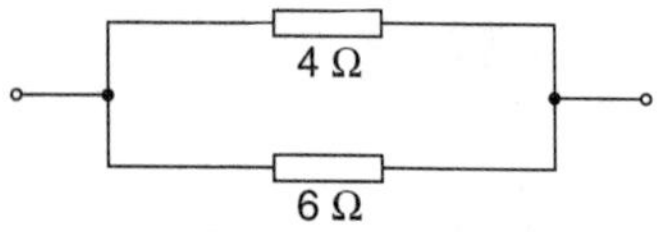

Figure 2.3 A parallel circuit. The voltage across each resistor is equal

There are two important points to note:

(a) The sum of the currents in the resistors is equal to the current flowing into the network.
(b) The p.d. is the same across each of the resistors.

Using these facts and a statement of Ohm's law ($I = V/R$) the proof may be shown. However, all that is needed is the following definition:

> For a parallel network of resistors, the reciprocal of the total effective resistance is equal to the sum of the reciprocals of the individual resistors.

This is shown mathematically as follows:

$$\frac{1}{R_{tot}} = \frac{1}{R_1} + \frac{1}{R_2} + \frac{1}{R_3}$$

Example 2.1

Two resistors of value 4 Ω and 6 Ω are connected in parallel as shown in Figure 2.3. Calculate the total effective resistance. (This is the same as saying what would be the value of the single resistor which could replace the 4 Ω and 6 Ω in parallel.)

Solution

Using the formula: $1/R_{tot} = 1/R_1 + 1/R_2$

$1/R_{tot} = 1/4 + 1/6$

Convert all the values into twelfths:

$$= \frac{3+2}{12} = \frac{5}{12}$$

therefore: $R_{tot} = 2.4\ \Omega$

Remember particularly that the quotient 5/12 is equal *not* to R_{tot} but the reciprocal of it, i.e. $1/R_{tot}$, so the total itself is equal to 12/5 which is 2.4 Ω.

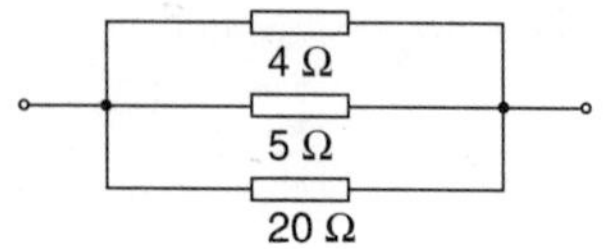

Figure 2.4 Three resistors in parallel

Example 2.2

Three resistors, each of value 4 Ω, 5 Ω and 20 Ω, are connected in parallel as shown in Figure 2.4. Calculate the total effective resistance of the network.

Solution

$$\frac{1}{R_{tot}} = \frac{1}{R_1} + \frac{1}{R_2} + \frac{1}{R_3}$$

$$\text{so:} \quad \frac{1}{R_{tot}} = \frac{1}{4} + \frac{1}{5} + \frac{1}{20}$$

$$\frac{1}{R_{tot}} = \frac{5+4+1}{20} = \frac{10}{20}$$

$$\text{and:} \quad R_{tot} = 2\,\Omega$$

Rules of thumb

We now look at four useful rules concerning parallel resistance which makes it much easier to solve resistive networks under some circumstances.

Rule 1 If there are only two resistors in parallel, then the total effective resistance may be calculated from:

$$R_{tot} = \frac{\text{Product}}{\text{Sum}}$$

Example 2.3

In Figure 2.5 we show 12 Ω and 4 Ω resistors in parallel. The total effective resistance can be calculated in the manner previously described as follows:

$$\frac{1}{R_{tot}} = \frac{1}{12} + \frac{1}{4} = \frac{1+3}{12} = \frac{4}{12}$$

$$\text{Hence:} \quad R_{tot} = 3\,\Omega$$

Using the 'product over sum' rule:

$$R_{tot} = \frac{12 \times 4}{12+4} = \frac{48}{16} = 3\,\Omega$$

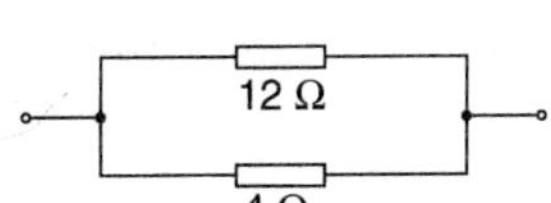

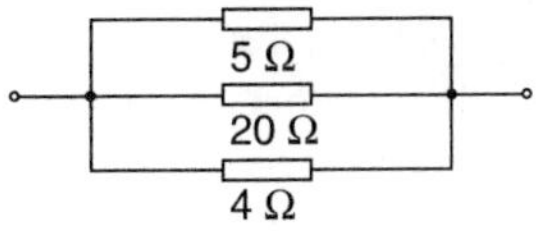

Figure 2.5 With only two resistors in parallel, the 'product over sum' rule may be applied

Figure 2.6 Using the 'product over sum' rule with three resistors in parallel

Rule 1 can *only* be used for a *pair* of resistors, but more complicated networks can be solved by taking the resistors in pairs. Noting the network shown in Figure 2.6 the following example should make this clear.

First take the 5 Ω and 20 Ω resistors in parallel:

$$R_{tot} = \frac{\text{Product}}{\text{Sum}} = \frac{5 \times 20}{5+20} = 4\,\Omega$$

Now take this total (4 Ω) in parallel with the remaining 4 Ω:

$$R_{tot} = \frac{\text{Product}}{\text{Sum}} = \frac{4 \times 4}{4+4} = 2\,\Omega$$

Note, incidentally, that when two resistors of the same value are placed in parallel, the total effective resistance is simply half of either, i.e. R_{tot} = half of 4 Ω = 2 Ω.

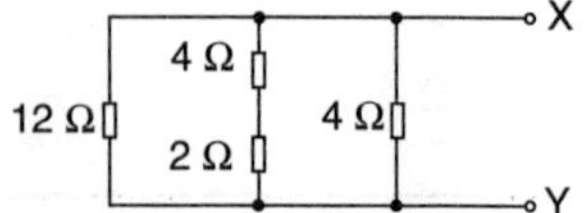

Figure 2.7 In a circuit like this, the total effective resistance is always less than the smallest

Rule 2 No matter how many resistors are placed in parallel, the total effective resistance will always be less than the smallest.

> Example 2.4
>
> Note the network shown in Figure 2.7.
>
> In Figure 2.7, the resistance measured between points X and Y would be: (a) 2 Ω, (b) 4 Ω, (c) 8 Ω, (d) 22 Ω.

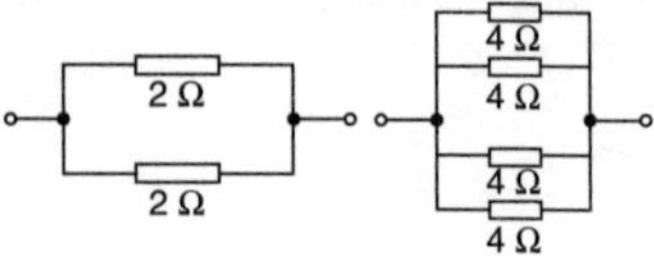

Figure 2.8 If the number of resistors is the same as the value of each, then R_{tot} = 1 Ω

Using Rule 2, the answer is obviously (a) without the need for calculation. (Since that's the only response which has a value less than the smallest.)

Rule 3 If the number of resistors in the network is the same as the value of each, then R_{tot} = 1 Ω.

By studying the two examples given in Figure 2.8, the validity of Rule 3 can soon be tested.

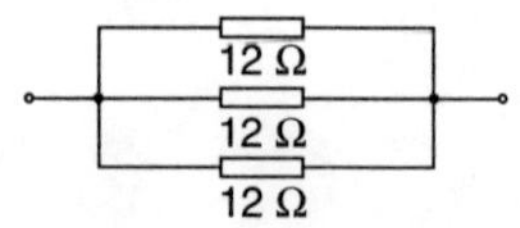

Figure 2.9 Here, R_{tot} is equal to the value of one of the resistors, divided by the number there are

Rule 4 Where several resistors in parallel all have the same value, divide the value of one of the resistors by the total number in the network.

> Examples 2.5 and 2.6
>
> Figures 2.9 and 2.10 give two examples. Work out R_{tot} for each one by any method you find the easiest and you will find that in Figure 2.9, R_{tot} = 12/3 = 4 Ω, and in Figure 2.10, R_{tot} = 100/4 = 25 Ω.

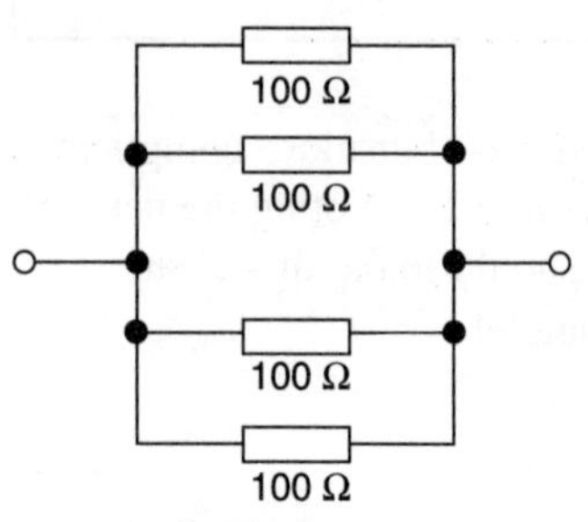

Figure 2.10 Another example of the application of Rule 4

> Exercise
>
> How many different ways (work out the values) can five 1 Ω resistors be connected in series, parallel or a combination of the two? Obviously, all five resistors can be connected in series giving 5 Ω, and in parallel giving 0.2 Ω. There are at least another twenty combinations. Some of the values of R_{tot} are given in the answers section.

Ohm's law calculations

The circuit shown in Figure 2.11 shows a network of series/parallel resistors. Supposing we wish to calculate the value of the current flowing through the 24 Ω resistor. Study the circuit carefully and then work through the questions given below to show how this may be done.

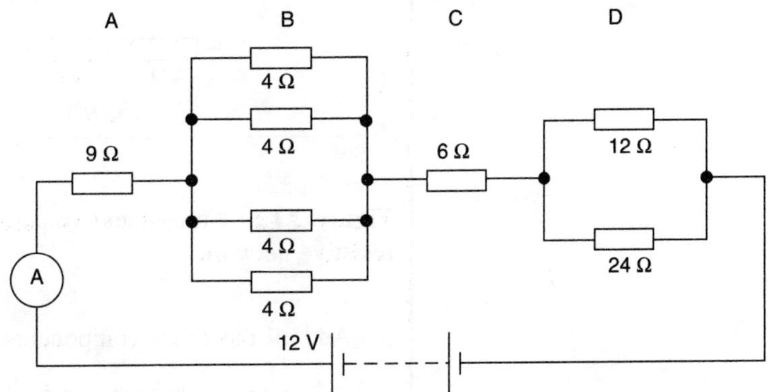

Figure 2.11 Current and voltage calculations for a resistive network

1. Calculate the total effective resistance of the parallel resistors in section B.
2. Calculate the total effective resistance of the parallel resistors in section D.
3. Calculate the total resistance in the circuit.
4. Given that the battery has a potential difference of 12 V, calculate the current flowing in the circuit.
5. Calculate the potential difference across the two resistors in section D.
6. Calculate the current flowing through the 24 Ω resistor.

You should have a value of 0.167 A. If you haven't, or are still unsure how to make the calculations, study the next example which gives all the necessary working out as well.

Example 2.7

In the circuit shown in Figure 2.12, calculate the current flowing through the 2 Ω resistor.

1. Calculate the total effective resistance of the first parallel section:

$$\text{i.e. } \frac{1}{R_{\text{tot}}} = \frac{1}{6+2} + \frac{1}{8} + \frac{1}{4+16} \; : \; R_{\text{tot}} = 3.33\ \Omega$$

2. Calculate the total effective resistance of the second parallel section, perhaps using the 'product over sum' rule:

$$\text{i.e. } R_{\text{tot}} = \frac{\text{Product}}{\text{Sum}} = \frac{10 \times 5}{10 + 5} = 3.33\ \Omega$$

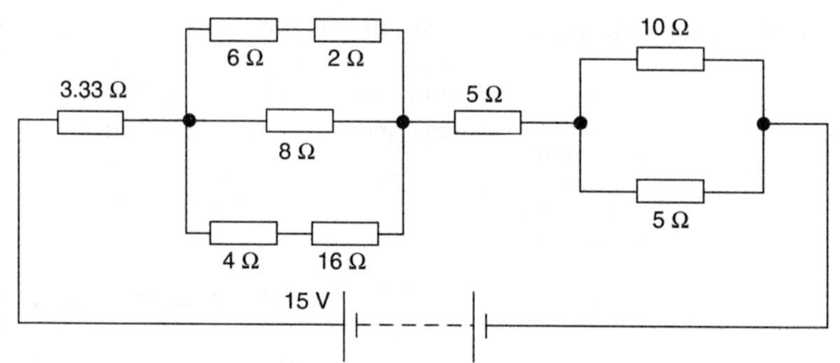

Figure 2.12 Current and voltage calculations for a different resistive network

3 Add all the series components together:

i.e. $3.33\,\Omega + 3.33\,\Omega + 5\,\Omega + 3.33\,\Omega = 15\,\Omega$

4 Having found the total effective resistance, and knowing the voltage, it is now possible to calculate the total current flowing from Ohm's law:

i.e. $I = V/R$ so: $I = 15/15 = 1\,\text{A}$

5 Calculate the potential difference (voltage) across the first parallel section (which contains the 2 Ω resistor of interest):

i.e. $V = I \times R = 1 \times 3.33 = 3.3\,\text{V}$

6 Calculate the current flowing through the 6 Ω and 2 Ω resistors in series (the current will be the same through each):

i.e. $I = V/R$ where $R = 6 + 2 = 8\,\Omega$

$I = 3.33/8 = 0.41625\,\text{A} = 0.42\,\text{A}$ (approx.)

This series of steps can be applied to almost any series/parallel network.

Potential dividers

Potential dividers can provide any fraction of a fixed potential. Figure 2.13 shows a simple potential divider network where V_0 is the total supply voltage and V_1 is the p.d. across R_1 which we can assume is the 'output voltage'.

In Figure 2.13,

$$I = V/R \text{ or } I = \frac{V_0}{R_1 + R_2}$$

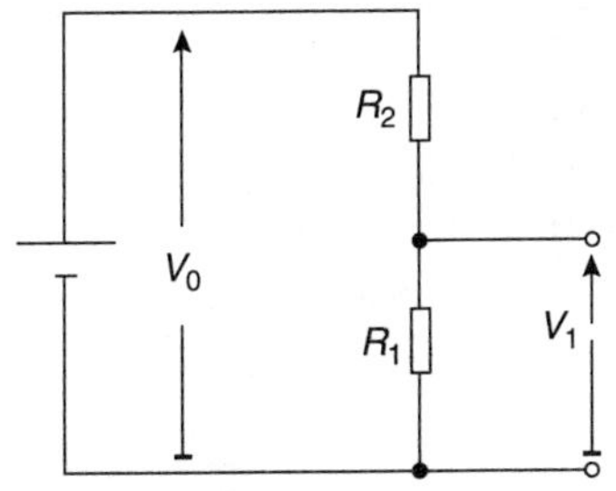

Figure 2.13 A potential divider network

The two resistors, R_1 and R_2 are in series, so the current flowing through each one *is the same*. Hence:

$$V_1 = IR_1$$

However, we have just stated that $I = V_0/(R_1 + R_2)$ and the value of I is the same, so we can say:

$$V_1 = \left[\frac{V_0}{R_1 + R_2}\right] R_1$$

By rearranging the final equation, we obtain:

$$V_1 = \left[\frac{R_1}{R_1 + R_2}\right] V_0$$

In other words, the potential difference across R_1 can be regarded as a fraction of the total voltage, V_0, as R_1 is a fraction of the total resistance, R_{tot} (where R_{tot} is equal to $R_1 + R_2$).

Example 2.8

If $V_0 = 6\,\text{V}$ and $R_1 = R_2 = 10\,\Omega$ calculate the p.d. across R_1 (i.e. V_1).

Solution

Using Ohm's law: $$I = \frac{V_0}{R_1 + R_2}$$

$$= \frac{6}{20} = 0.3\,\text{A}$$

Hence: $$V_1 = 0.3 \times 10 = 3\,\text{V}$$

Think carefully about the solution; all we have done is to use Ohm's law. Now, knowing that in a series circuit the current is the same, the current through both R_1 and R_2 is therefore the same and the value of each resistor is the same ($10\,\Omega$) so ... yes! the voltage dropped across each resistor is also the same. If the supply is 6 V, then the only solution which allows both resistors to have the same voltage drop across them is 3 V.

Going one stage further, does it matter what the value of the resistors in Figure 2.13 is? The answer is no, because if the resistance value is increased, less current flows, but it is still the same current through each resistor and therefore the same voltage is dropped across each. As long as R_1 and R_2 have the same value, V_1 will always be half the supply voltage.

Example 2.9

In Figure 2.13, the supply voltage remains the same but the resistor values now become 1000 Ω (1 kΩ). Calculate the voltage across R_1.

Solution

$$V_1 = \frac{6}{1000 + 1000} 1000$$

$$\text{or:} \quad V_1 = \frac{1000}{1000 + 1000} 6$$

$$\text{hence:} \quad V_1 = \frac{1}{2} \times 6 = 3\,\text{V}$$

This example shows that we obtain the answer of 3 V as we expected, but it also shows that it is simply half of the supply voltage. And it doesn't matter what the value of the resistors is, as long as they have the same value. In fact it's the *ratio* between the two values that's important, not the actual values.

Example 2.10

Suppose a 9 V supply is available and it is required to produce 3 V, 4.5 V and 6 V using 3 pairs of resistors in a potential divider configuration. Work out the values for the potential dividers, assuming that 1/2 W resistors are available.

Solution

It is clear that any number of resistor values would produce these voltages as long as they are in the correct proportions. However, in order to ensure adequate current rating, we first need to work out the minimum resistance required in the circuit.

$\frac{1}{2}$ W resistors are used and power = V^2/R

$$\text{hence:} \quad R_{tot} = \frac{9^2}{0.5} = \frac{81}{0.5} = 162\,\Omega$$

Using a larger value would ensure that the current is safely limited, so let's suggest 220 Ω. Using the potential divider equation, where $V_0 = 9\,\text{V}$, $V_1 = 3\,\text{V}$ and $R_1 + R_2 = 220\,\Omega$:

$$V_1 = \frac{V_0}{R_1 + R_2} R_1 \quad \text{rearranging: } R_1 = \frac{V_1(R_1 + R_2)}{V_0}$$

$$\text{hence:} \quad R_1 = \frac{3 \times 220}{9} = 73\,\Omega \text{ (would use } 75\,\Omega\text{)}$$

The value of R_2 is therefore 220 Ω − 73 Ω = 147 Ω so a preferred value of 150 Ω would probably be used. In order to obtain 4.5 V (half V_0) $R_1 = R_2$ so $100\,\Omega$ would be adequate. For 6 V, $R_1 = 75\,\Omega$ and $R_2 = 150\,\Omega$, which can easily be verified, but which is obvious from previous calculations.

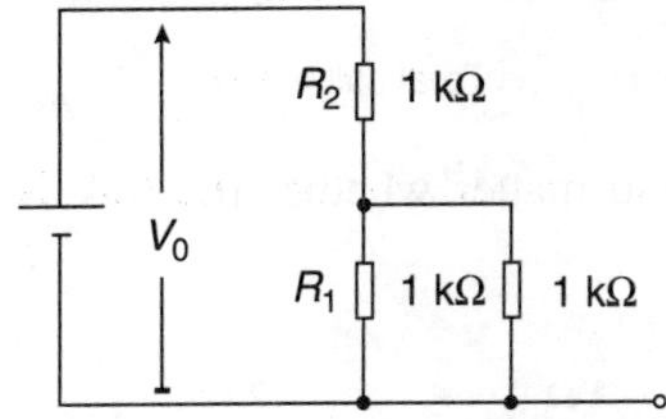

Figure 2.14 A loaded potential divider

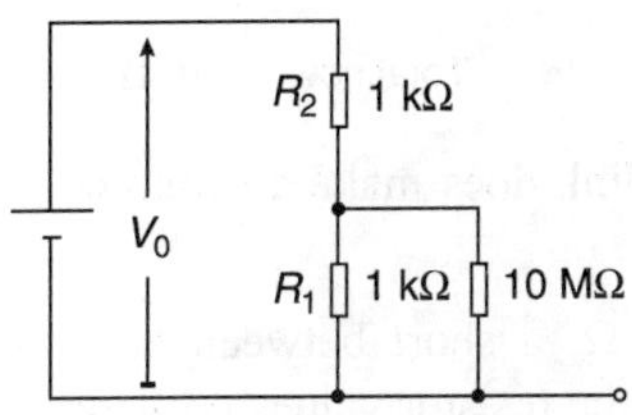

Figure 2.15 Measuring the output voltage with a DVM

Although the theoretical values are quite correct, in practice, what voltage actually appears across R_1 will depend upon what it is connected to. Even connecting a meter across R_1 will change the voltage because all meters have internal resistance. What this means is that when the meter is connected, it's like putting another resistor in parallel, the total effective resistance then reduces, and the voltage drop at the output reduces. If the 'internal resistance' of the meter has a value of $1\,k\Omega$ and we use two $1\,k\Omega$ resistors for the potential divider, the circuit looks like that shown in Figure 2.14.

The two $1\,k\Omega$ resistors in parallel represent R_1 and the internal resistance of the meter, respectively. Two $1\,k\Omega$ resistors in parallel have a total effective resistance of only $500\,\Omega$. R_{tot} is therefore $1500\,\Omega$ ($1.5\,k\Omega$) and the current becomes $I = V/R$ which is $6/1500 = 4\,mA$ (0.004 A). Further calculation reveals that the p.d. across R_2 has risen to 4 V while that across R_1 has fallen to only 2 V. Fortunately, most meters used by technicians and engineers have internal resistances far greater than $1\,k\Omega$; digital voltmeters, for example, have internal resistances of at least $10\,M\Omega$, and in many cases even more.

If a meter with an internal resistance of $10\,M\Omega$ is used, the circuit can be represented by Figure 2.15.

Calculation then shows that the connection of the meter across R_1 makes hardly any difference since in this condition the voltage becomes 2.998 V – a difference which is negligible.

The Wheatstone bridge

City and Guilds 224 Part 1 students are not required to study this section even though its understanding relies only on a good working knowledge of Ohm's law, and a bit of maths. However, it is included as another example of the behaviour of current and voltage in resistive networks and it is useful to understand the basic underlying principles of the Wheatstone bridge now, if at all possible, since it will reappear in higher level work.

Figure 2.16 shows two simple networks consisting only of $1\,\Omega$ resistors. On the left, a simple calculation reveals that the total effective resistance is $1\,\Omega$. Now, what happens if the junctions between each pair of resistors are connected together? The network becomes like the one on the right.

Again, a simple calculation reveals that R_{tot} is $1\,\Omega$. So, the value of R_{tot} is $1\,\Omega$ when there is an infinite resistance between the junctions and it's just the same if there's a link – a short circuit (S/C) – between them.

Now take a look at Figure 2.17. One of the resistors has been changed from $1\,\Omega$ to $2\,\Omega$. Calculation of R_{tot} produces a value of $1.2\,\Omega$; with a link between the two junctions a further calculation shows that R_{tot} has become $1.16\,\Omega$.

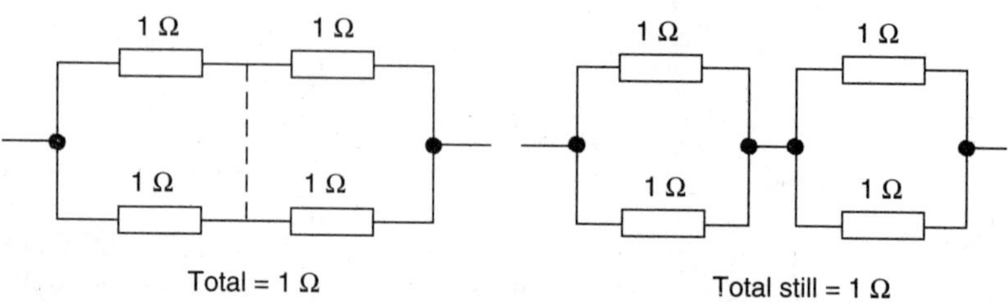

Figure 2.16 In this case, it doesn't seem to matter whether the link is there or not

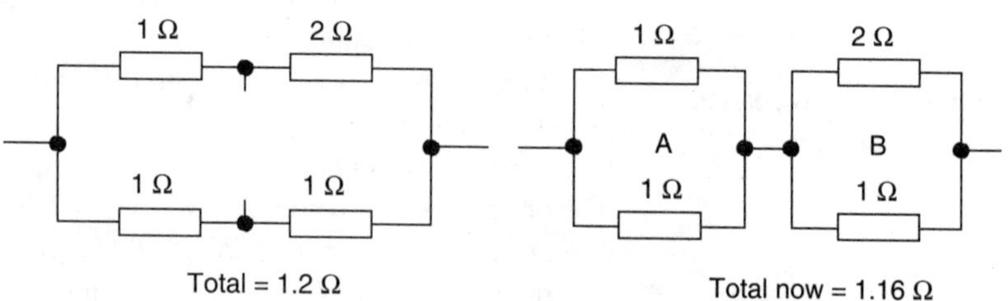

Figure 2.17 In this case, inclusion of the link does make a difference

So we see that if the resistors are all 1 Ω, a short between the two junctions makes no difference. When one of the resistor values is changed, there is a difference. The reason is that if all the resistors have a value of 1 Ω *no current flows* between the junctions, so it doesn't matter what is placed between them, an infinite resistance (no short) or zero resistance (a short circuit) or, in fact, any other value in between. When a different value is introduced, the imbalance causes a current to flow between the junctions so the voltage dropped across each half of the network is different. The resistors do not even have to be all the same value, they simply have to be in the correct ratios. This is the basis of the Wheatstone bridge. When the bridge is balanced, no current flows throw the central resistance.

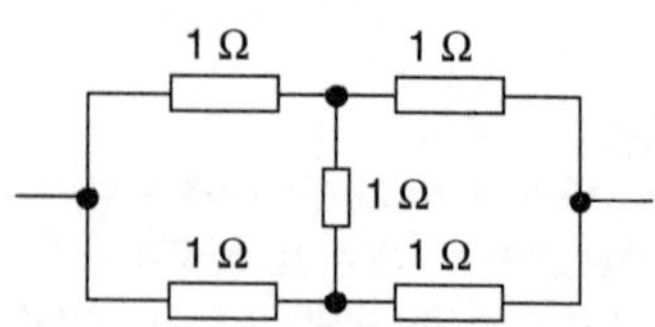

Figure 2.18 One configuration of five 1 Ω resistors

Earlier on in this chapter we described an exercise in which many different combinations of five 1 Ω resistors were asked for. If a 1 Ω resistance were to be placed between the junctions of the resistors in Figure 2.16 it becomes like the circuit shown in Figure 2.18.

From recent discussions you should now see that R_{tot} for the Figure 2.18 network is 1 Ω. The central 1 Ω resistance has no effect, it could be any value from zero to infinity and it would still make no difference. This is a useful point to ponder. To solve networks which have unbalanced values requires the use of Kirchhoff's laws, which are not required at this level.

Conductors, semiconductors and insulators

Figure 2.19 shows the 'conductor continuum'. In this context, a continuum is simply a straight line along which we place the data that are of interest. For example, we could have a continuum which showed the heights of people in this country. At one end, there would be the numbers of very short people, at the other end there would be all the very tall people and in the centre would be most of the people who are of 'average' height.

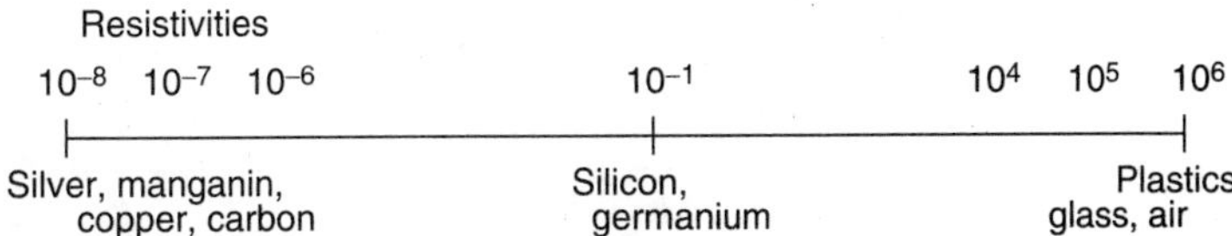

Figure 2.19 The conductor continuum. Good conductors on the left, poor conductors (insulators) on the far right, semiconductors in the middle. Just a few examples are given here

A continuum could be used to represent all kinds of data; in this case we are looking at conductors of electricity. So at one end we have the best conductors, at the other end the very poor conductors (also called insulators) and in the middle one or two semiconductors.

One of the best conductors of electricity is silver (Ag), but it is also very expensive. Copper (Cu) is also an excellent conductor of electricity. Although it's not quite as good as silver, it's a whole lot cheaper, and so is in very common use in electrical and electronic wiring being a good compromise between cost and conductivity.

Metals in common use in electrical and electronic work include: copper (wiring), aluminium (chassis, etc.), brass (terminals), silver (specialised wiring requirements), iron (in the cores of transformers, etc.) and tungsten. The latter is used for lamp filaments because it melts at 3380°C. Although gold does not have as low a resistance as silver, it is used for interconnections in integrated circuits because of its resistance to corrosion. It is unaffected by air (does not tarnish), water, sulphur dioxide and most acids.

To make wirewound resistances, an alloy called manganin is often used. This is made from copper (Cu), manganese (Mn) and nickel (Ni). Nichrome (Ni–Cu–Cr) is used for electric fires – it does not oxidise at 1000°C. Carbon (graphite) is used for smaller resistors in electronic work.

Semiconductors

The basic semiconductor materials are silicon, germanium and selenium. By far the most important is silicon, since most transistors and integrated circuits are made from it. Many semiconductor alloys have recently been produced which are used in the manufacture of light-emitting diodes (LEDs) and other electronic components.

Insulators

In the past, wood and bakelite were used as insulating materials in electrical work but these have now been largely superseded by plastics. All plastics are good insulators and some are superb. Other insulators include ceramics, glass, PVC, paper, mica and polystyrene. Many of the latter are used as the insulating material (dielectric) in capacitors. The common uses of some conductors and insulators are listed in Table 2.1.

Table 2.1 Some uses of conductors and insulators

GOOD CONDUCTORS		INSULATORS	
Metal	*Use*	*Insulator*	*Use*
Aluminium	Chassis	Air	Between 13 A plug pins!
Brass	Terminals	Aluminium oxide	Dielectric
Carbon (C)	Batteries	Bakelite	Old radios
Copper (Cu)	Wires	Ceramic	Dielectric
Germanium	Transistors	Glass	Light bulbs
Gold (Au)	IC terminals	Mica	Dielectric
Iron	Transformers	Paper	Dielectric
Lead	Car batteries	Polycarbonate	Dielectric
Magnesium	Flash bulbs	Polystyrene	Dielectric
Mercury	Tilt switches	Porcelain	Insulators
Nichrome	Wirewound resistors	PVC	Dielectric
Nickel	Rechargeable batteries		
Silicon (Si)	Transistors		
Silver (Ag)	Special plugs		

Atomic structure

We saw in Chapter 1 how metals with 'free' electrons were good conductors, because an electric current consists of an orderly drift of electrons in a particular direction. But some elements and compounds combine together in different ways.

In insulators, for example, there are no free electrons so the material doesn't conduct unless you raise the potential difference to thousands, or even millions, of volts. In between these two extremes are special elements called semiconductors. We've already mentioned silicon as an important example; this element combines with itself to form what are called 'covalent' bonds. Copper will conduct if a potential difference of only 1 μV (a millionth of a volt) is placed across it; some insulators won't conduct even if a potential of 1 MV (one megavolt or a million volts) is applied; semiconductors, however, need only of the order of 1 V in order to make them conduct. The potential difference applied breaks the covalent bonds allowing some electrons to move into the conduction band and take part in an electric current.

Resistance

The resistive nature of a conductor determines how much it will resist an electric current. If a current is 'resisted' by a resistor, a voltage is dropped across it and energy (heat) is dissipated. Hence, good conductors have a very low resistance and insulators have a very high resistance.

Supposing we compare a wire made from silver with a wire made from copper. Which wire has the least resistance? One immediate response might be that the silver has the least resistance since we have most recently stated that silver is one of the best conductors known, and we only use copper because it is a lot cheaper. In answer to this, however, we would have to say what about the thickness or cross-sectional area (c.s.a.) of the wire – doesn't that make a difference? The fact is, of course, that it does. So which has most resistance, a thick wire or a thin wire?

While you're thinking about it, what diameter pipe carries the most water, 10 cm or 20 cm? Obviously the larger diameter pipe – the 20 cm one – carries the most water. So the answer to the wire problem must be that the wire with the largest c.s.a. will carry the most current and therefore has the least resistance.

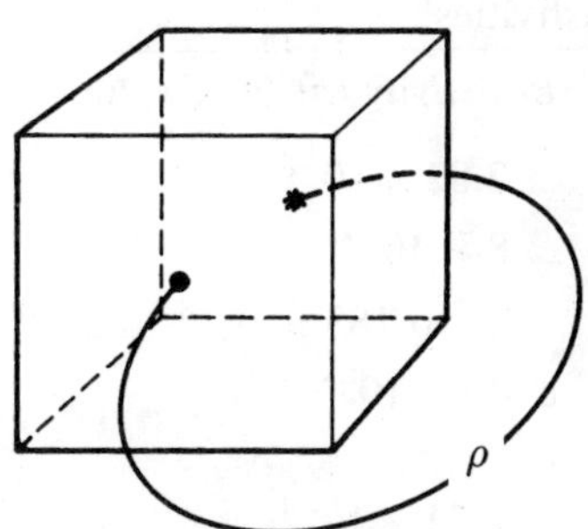

Figure 2.20 The resistivity of a material is the resistance between opposite faces of a cube of the material, the sides of which are 1 metre long

So, in order to determine how well a conductor resists a current (or how well it conducts) we need to know what it is made of, how long it is and what is its c.s.a. What it is made of is determined by its resistivity (given the Greek symbol ρ, rho – the same as density), its length, l (in metres) and its c.s.a. in m^2. Combining these three quantities together, the resistance, R, can be calculated from:

$$R = \frac{\rho l}{A}\ \Omega\text{m}$$

The units of resistivity are ohm-metres (Ωm), that is to say, ohms times metres, *not* ohms per metre. Resistivity can be defined as: the resistance between opposite faces of a cube of the material, the sides of which are 1 metre long as shown in Figure 2.20.

The next question is to ask what sort of resistance value you would expect to read across opposite ends of a cube of the material. Thousands of ohms? A few ohms? A few millionths of an ohm? In fact the latter is most correct; the resistance will be tiny since the c.s.a. is so large. If the material is copper, then the value would be about $0.017 \times 10^{-6}\ \Omega$m. It is this value which is known as the resistivity (formerly specific resistance).

Example 2.11

What is the resistance of 3 m of nichrome wire, c.s.a. $= 2 \times 10^{-7}\ \text{m}^2$ and resistivity $1.1 \times 10^{-6}\ \Omega$m?

Solution

$$R = \frac{\rho l}{A} = \frac{1.1 \times 10^{-6} \times 3}{2 \times 10^{-7}} = 16.5\ \Omega$$

Example 2.12

What is the resistance of 3 m of copper wire, c.s.a. same as in the last example (2×10^{-7} m^2), the value of resistivity being 1.72×10^{-8} Ωm?

Solution

$$R = \frac{\rho l}{A} = \frac{1.72 \times 10^{-8} \times 3}{2 \times 10^{-7}} = 0.258\ \Omega$$

This is only a little over $\frac{1}{4}$ Ω!

Some resistivity values are given in Table 2.2.

Table 2.2 Some metals and their resistivities

Metal	*Resistivity (at* 20°C) Ωm
Aluminium	2.82×10^{-8}
Brass	c. 8×10^{-8}
Carbon (graphite)	33 to 185×10^{-8}
Constantan (also called Eureka: 60% Cu, 40% Ni)	c. 49×10^{-8}
Copper	1.72×10^{-8}
Iron	c. 9.8×10^{-8}
Manganin (used for wirewound resistors: 84% Cu, 12% Mn and 4% Ni)	c. 44×10^{-8}
Mercury	95.77×10^{-8}
Nichrome (used for electric fires – does not oxidise at 1000°C; Ni–Cu–Cr)	c. 100×10^{-8}
Silver	1.62×10^{-8}
Tungsten (used for lamp filaments – melts at 3380°)	5.5×10^{-8}

Questions

1 Calculate the total effective resistance of: 1 Ω, 10 Ω, 100 Ω and 1000 Ω resistors in series.

2 Calculate the total effective resistance of a 15 Ω resistor in parallel with a 25 Ω resistor.

3 12 Ω and 8 Ω resistors are placed in parallel and are then wired in series with 22 Ω, 2.2 Ω and 11 Ω resistors as shown in Figure 2.21. Calculate the total effective resistance of the network.

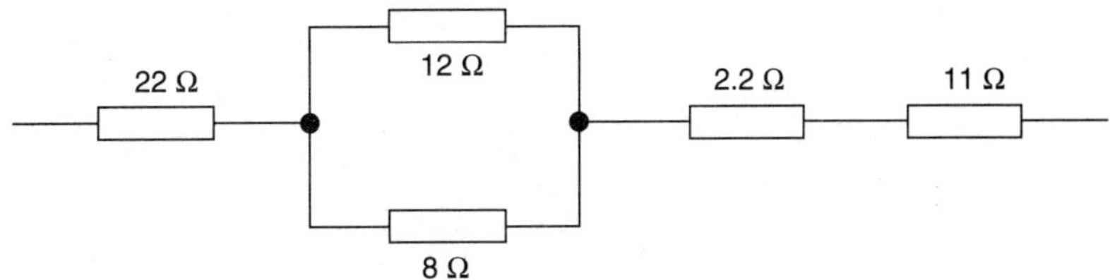

Figure 2.21

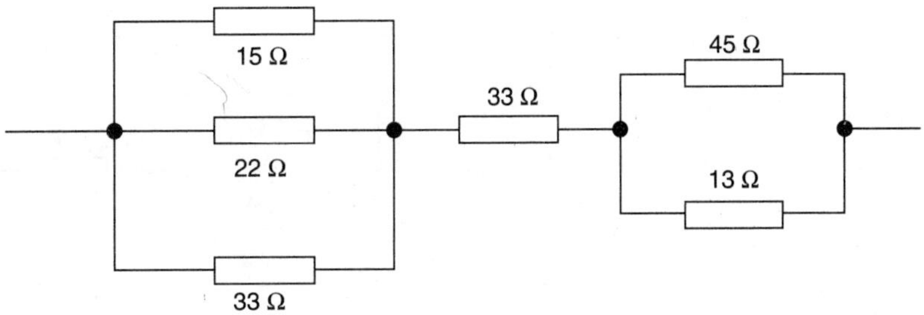

Figure 2.22

4 Calculate the total effective resistance of the network shown in Figure 2.22.

5 Calculate the total effective resistance of the network shown in Figure 2.23.

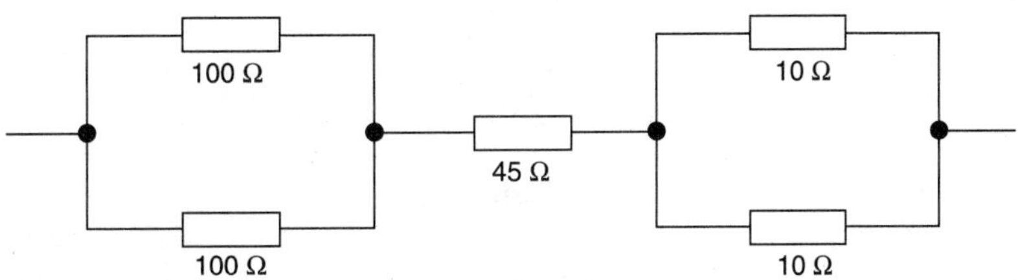

Figure 2.23

6 Calculate the total current flowing in the circuit of Figure 2.24.

7 If 50 V is applied across x and y in the circuit shown in Figure 2.25, find the current in the 5 Ω resistor.

8 In the circuit of Figure 2.26, a current of 5 A enters at C and leaves at B. Calculate the p.d. between C and A.

9 In the circuit of Figure 2.27, calculate the current flowing through the 4 Ω resistance marked B.

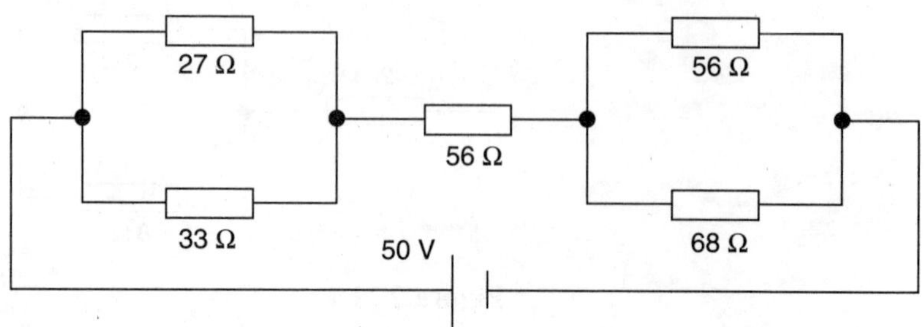

Figure 2.24

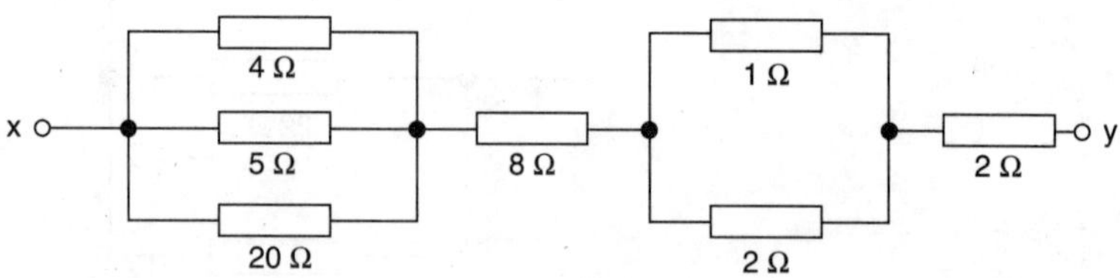

Figure 2.25

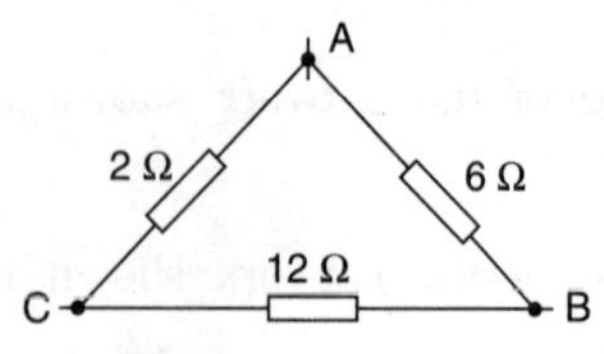

Figure 2.26

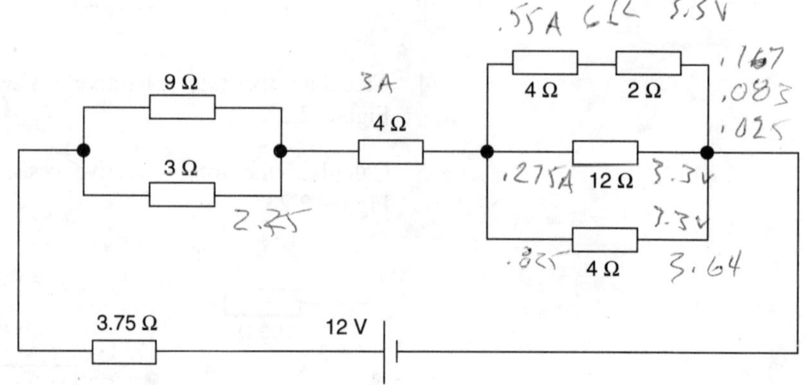

Figure 2.27

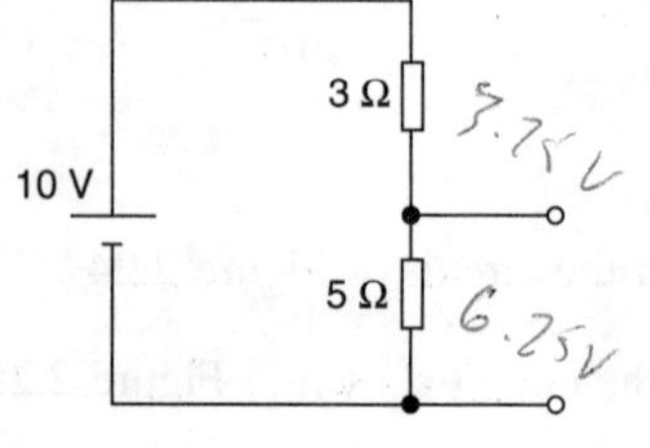

Figure 2.28

10 A 5 Ω and 3 Ω resistor are connected in series as shown in Figure 2.28. A 10 V supply is connected across both resistors. Determine the voltage across each resistor.

11 A 6 Ω and 66 Ω resistor are connected in series to a 12 V supply in a similar manner to that shown in Figure 2.28. Determine the voltage across the 6 Ω resistor.

12 A 30 kΩ resistance is in series with a 40 kΩ resistor and connected to a 100 V supply. A meter is placed across the 40 kΩ resistor. If the internal resistance of the meter is also 40 kΩ what voltage will it measure?

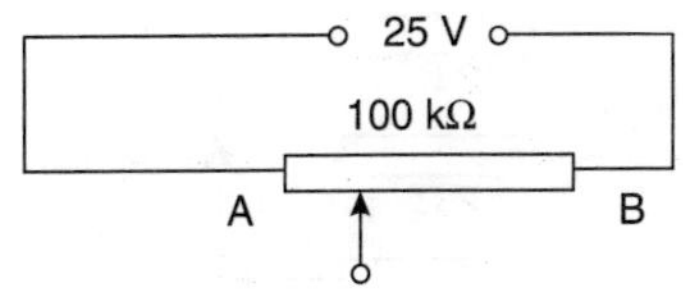

Figure 2.29

13 A variable potentiometer of 100 kΩ is placed across a 25 V supply. The wiper is 1/5th the distance along its track as shown in Figure 2.29. What is the p.d. between the wiper in this position and the point B?

14 Name the particles of an atom which have a negative charge.

15 From the following list – aluminium, copper, glass, gold, iron, mica, polystyrene, brass, silicon, mica, paper, PVC (polyvinyl chloride), magnesium, mercury, air and nichrome – name one material in each case which is used as:

(a) the conductor in the connecting wires used in house wiring
(b) the semiconductor used in the manufacture of transistors
(c) the pins and terminals of a 13 A plug
(d) the insulating material in some electric cables
(e) the conductor that is used to connect an integrated circuit to the pins on the plastic case
(f) the dielectric in some capacitors
(g) the strong metalwork used for radio and TV chassis
(h) the conductor in tilt switches.

16 A manganin wire has a resistivity of 4.5×10^{-7} Ωm, a length of 2 m and a c.s.a. of 1.11×10^{-7} m^2. What is its resistance?

17 A Eureka wire has a resistance of 2 Ω. If its length is 1 m and c.s.a. 2.5×10^{-7} m^2, calculate its resistivity.

18 A resistance of 250 Ω is to be made from nichrome wire, c.s.a. 5.2×10^{-9} m^2. If the resistivity of nichrome is 1×10^{-6} Ωm, what length of wire would be required?

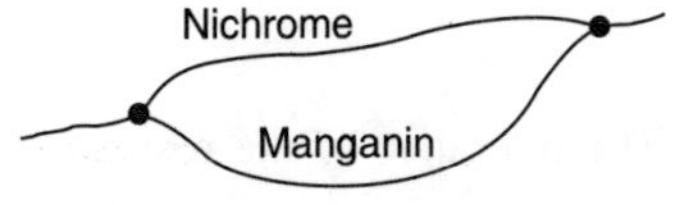

Figure 2.30 A 12 m pair of resistance wires connected in parallel

19 An electric fire element is to be made from nichrome wire whose resistivity is 10^{-6} Ωm. If the length must be 2 m and the resistance 80 Ω, what c.s.a. wire would you use?

20 Two wires 12 m long are connected together at each end as shown in Figure 2.30.

The nichrome wire has a resistivity of 1.06×10^{-6} Ωm and a diameter of 0.9 mm. The manganin wire has a resistivity of 0.42×10^{-6} Ωm. A p.d. of 40 V across the two wires produces a total current of 3 A. What is the diameter of the manganin wire?

(Hints and a solution to this question are provided in the answers section.)

Multiple choice questions

1 In Figure 2.31, if all the resistors have a value of 12 Ω, the total effective resistance of the network is

A 6 Ω
B 8 Ω

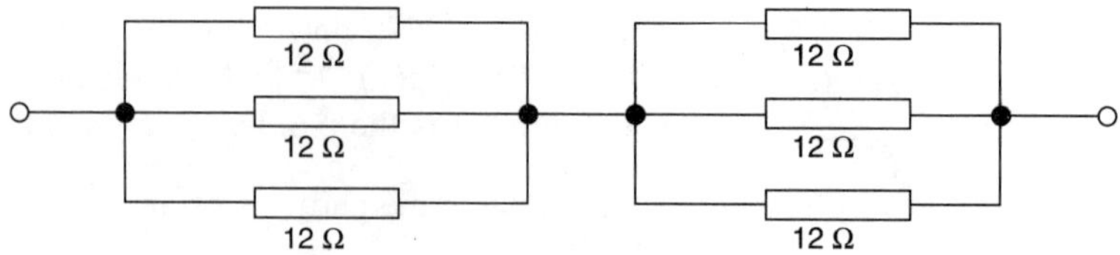

Figure 2.31

C 24 Ω
D 48 Ω

2 In Figure 2.32, the current supplied by the battery is

A 0.5 A
B 1.0 A
C 1.5 A
D 2.0 A

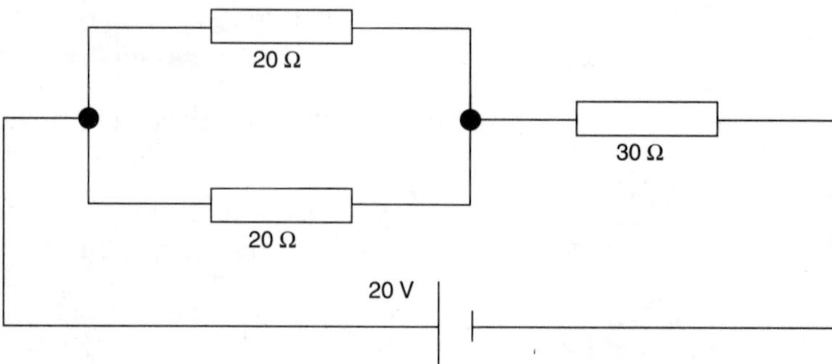

Figure 2.32

3 The fourth band on a resistor is gold. The tolerance of the resistor is

A ±1%
B ±2%
C ±5%
D ±10%

4 The resistance of a length of wire, l, and cross-sectional area, A, is proportional to

A $A \times l$
B l/A
C l/lA
D A/l

5 The SI unit for power is

A amperes
B newtons

C watts
D coulombs

6 A resistor with three bands coloured red, red, orange has a value of

A 2.2 kΩ
B 22 kΩ
C 220 kΩ
D 2.2 MΩ

7 The nearest E12 series preferred-value resistor to 80 Ω is

A 68 Ω
B 75 Ω
C 82 Ω
D 100 Ω

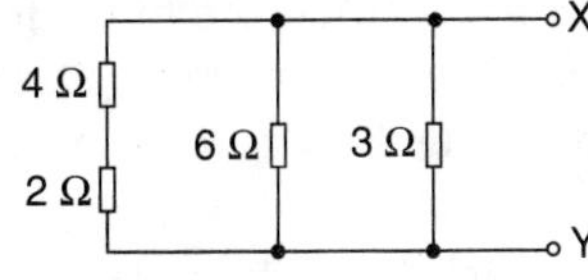

Figure 2.33

8 In Figure 2.33, the total effective resistance across X and Y is

A 1 Ω
B 1.5 Ω
C 3 Ω
D 15 Ω

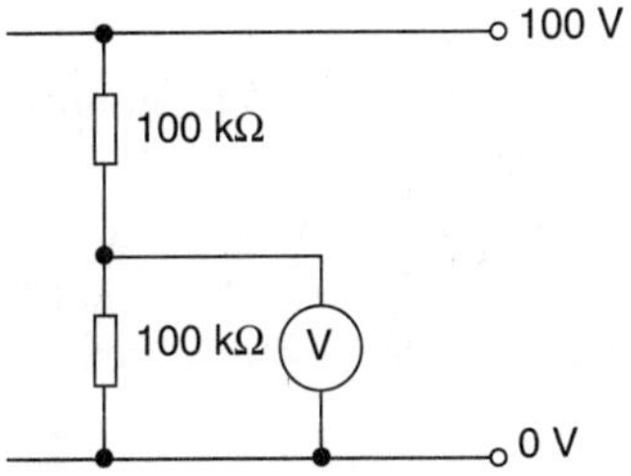

Figure 2.34

9 In the potential divider shown in Figure 2.34, if the voltmeter has an internal resistance of 100 kΩ, the meter reading will be approximately

A 33 V
B 50 V
C 66 V
D 100 V

10 The metal with the least resistivity in the list below is

A copper
B gold
C silver
D aluminium

3 Electrostatics and capacitance

I – Electrostatic charges

Capacitors are rather like small car batteries in that they can store a charge and then release it when connected into a circuit. However, the mechanism is quite different, car batteries producing a potential difference through a chemical reaction and capacitors storing charge by electrostatic attraction. Capacitors cannot *create* charge and they rapidly discharge when connected into a circuit.

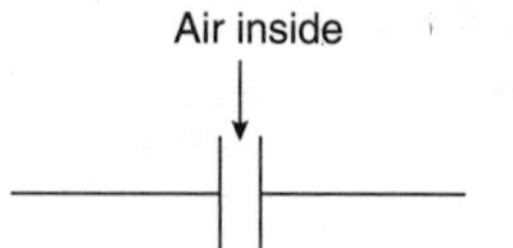

Figure 3.1 Basic capacitor

Basically, a capacitor consists of two metal plates brought very close together as shown in Figure 3.1.

This arrangement doesn't seem to do very much, but look what happens when it is connected into a circuit as shown in Figure 3.2. In this figure electrons flow into the capacitor and charge builds up on the plates as shown.

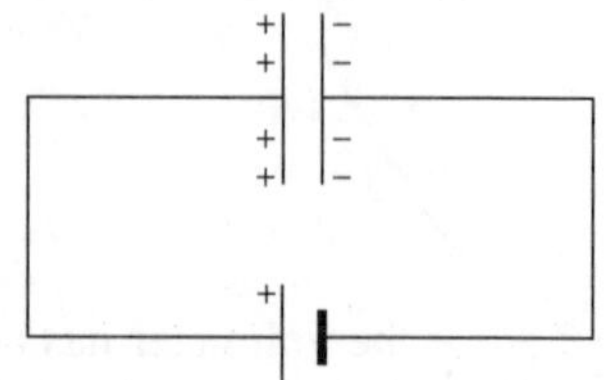

Figure 3.2 The capacitor being charged

The charge is maintained through electrostatic attraction. The opposite charges attract each other like the opposite poles of a magnet. We can assume that the two plates are separated by air. In fact, any number of insulating materials may be used, including paper, mica, polystyrene and ceramic. Whatever it is, the insulating material is called a dielectric.

Practical capacitors

Practical capacitors usually consist of two long strips of metal foil, separated by long strips of dielectric rolled up like a 'Swiss roll' as shown in Figure 3.3.

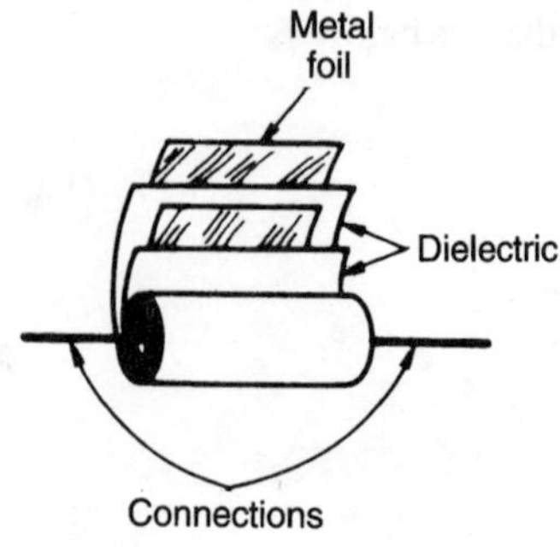

Figure 3.3 A practical capacitor

The arrangement allows plates of large area to be close together in a small volume. Plastics like *polyesters* are commonly used as the dielectric, with films of metal deposited on the plastic to act as the plates.

Electrolytic capacitors

In the *electrolytic* type a very thin dielectric of aluminium is formed chemically between the two strips of aluminium foil, giving a high value, compact capacitor. These devices are *polarised*, i.e. they have positive and negative terminals, and it is very important to take this into consideration when connecting an electrolytic capacitor into a circuit.

Circuit symbols

There are two basic symbols as shown in Figure 3.4. Two additional symbols are also given to cover variable capacitors. The preset variable is often also known as a 'trimmer'.

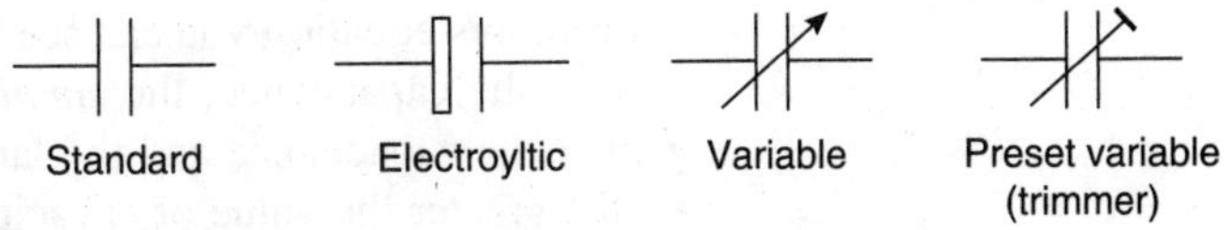

Figure 3.4 The four main capacitor symbols

Units

The unit of capacitance is the farad, F. It is a very large unit so that most practical capacitors have units of μF, nF and pF.

μ (micro, a millionth) $= 10^{-6}$ F

n (nano, a thousand millionth) $= 10^{-9}$ F

p (pico, a million millionth) $= 10^{-12}$ F

1 pF is sometimes known as 1 μμF. The unit mF (millifarad) is hardly ever used, in fact it has now become common to quote all values in nF so that a 0.1 μ capacitor would most frequently be referred to as 100 nF or, simply, 100 n. The largest practical capacitor is about 100 000 μF and the smallest about 0.5 pF. Very recently, capacitors with values as high as 3 F have been produced to act as back-up power supplies for computer memories. This is, however, a rather specialised application.

Generally, capacitor values of 1 μF or above are electrolytic types and those <1 μF are non-electrolytics. As in all things, however, there are always exceptions.

Equivalent capacitor values – some examples:

0.1 μF = 100 nF   0.01 μF = 10 nF

0.1 nF = 100 pF   10 000 pF = 10 nF

The capacitance of a capacitor depends only upon its physical construction, namely the area of the plates, A (m^2), the insulating material between the plates (called the 'dieletric'), ε_r (no units) and the distance between the plates, d (m). The value of capacitance increases as the plate area, A, increases, and when the value of ε_r increases. As far as the distance, d, between the plates is concerned, a reduction of this value produces an increase in capacitance. All of this is summarised in the box below:

> The area of the plates, A(m^2)
> The distance between the plates, d (m)
> What is between the plates, i.e. the dielectric, ε_r (no units)
> The permittivity of free space, ε_r (F/m)

Putting all these values together, we have:

$$C = \frac{\varepsilon_0 \varepsilon_r A}{d} \text{ farads}$$

From this equation you can see that the *larger* the plate area, the greater will be the capacitance; the *smaller* the distance between the plates, the greater the capacitance and the larger the value of the dielectric constant, ε_r, the greater the value of capacitance. The value ε_0 is called the Permittivity of Free Space:

It is equal to: $\varepsilon_0 = 8.85 \times 10^{-12}\,\text{Fm}^{-1}$

The permittivity of air has the same value to a close approximation; anything else placed between the plates must be taken account of by multiplying ε_0 by ε_r.

Types of capacitor

1 Aluminium electrolytic

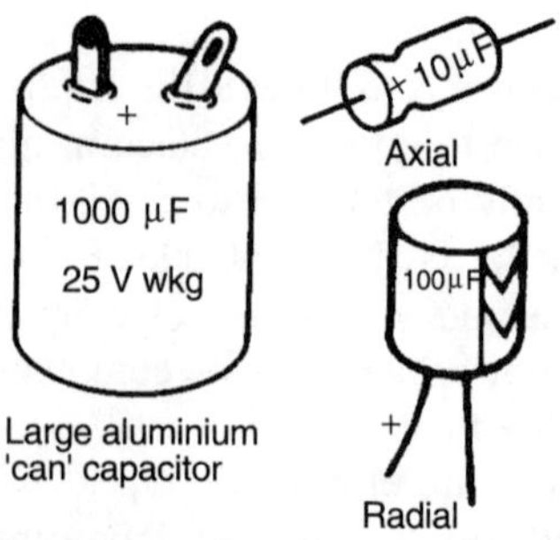

These have a high capacitance for a relatively small volume. Common values range from 1 µF to 10 000 µF. Electrolytics must be placed in a circuit the correct way round, i.e. you must observe the *polarity*. Applications include: smoothing, decoupling, AF coupling, photoflash charge storage. Some typical packages are shown in Figure 3.5.

+ 10µF
Axial
1000 µF
25 V wkg
100µF
Large aluminium 'can' capacitor
Radial

Figure 3.5 Electrolytics

2 Solid tantalum

Tantalum capacitors, like electrolytics, have a large capacitance for a small volume, and are polarised. Two basic types are illustrated in Figure 3.6. These capacitors have values between 0.1 µF and 100 µF and are generally used for smoothing and decoupling purposes. They are also used as AF coupling capacitors where space is at a premium.

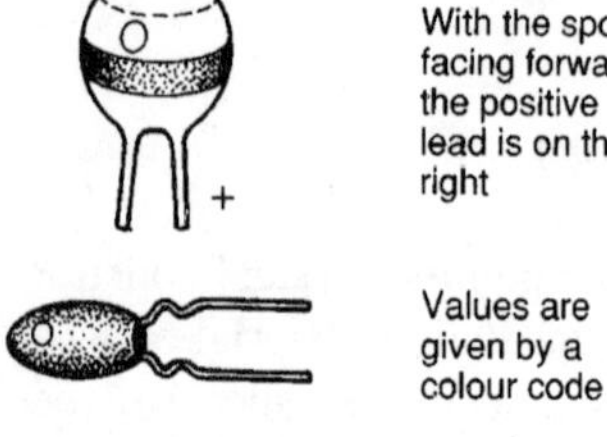
With the spot facing forward the positive lead is on the right
+
Values are given by a colour code

Figure 3.6 Tantalum capacitors

3 Ceramic

Most types have values printed on them. Some types have a colour code. This type of capacitor is not polarised, i.e. they do not have negative and positive leads, and it does not matter which way round in a circuit they are connected. Various ceramic capacitors are pictured in Figure 3.7.

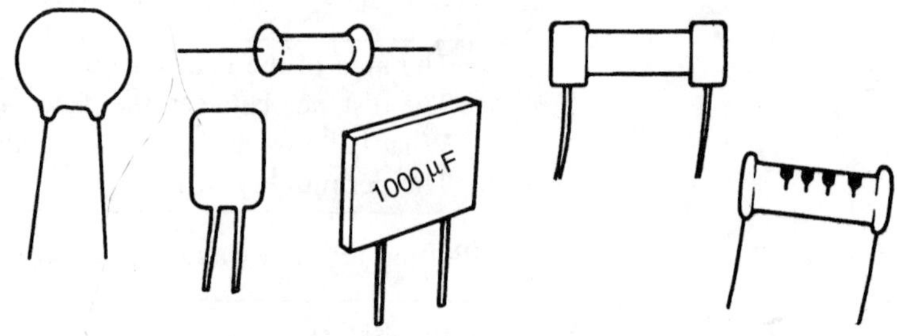
1000 µF

Figure 3.7 Various ceramic capacitors, including disc and plate types

Values Disc and plate types: 0.01 µF to 0.47 µF at 3 V to 500 V. Tubular types: 0.5 pF to 10 000 pF at 500 V.

Applications General purpose AF and RF. Very much used where space is at a premium. Coupling and decoupling. Certain types have calibrated temperature coefficients which may be used for compensation in temperature sensitive tuned circuits.

Working out the values Ceramic disc types often have the first two figures written on them and the third is the number of noughts, the unit being the pF. So, 472 would be 4700 pF or 4.7 nF. An example such as 470 means there are no noughts, so the value is 47 pF. It is also common to find the tolerance letter given, e.g. 102 H which gives 10 as the first two figures, 2 as the number of noughts and H as ±2.5% tolerance, so the capacitor is 1000 pF (or 1 nF) and has a tolerance of 2.5%.

4 Polyester and polycarbonate

Polycarbonate capacitors also have a colour code for their values. Applications include AF filtering, AF coupling and decoupling as well as many general purpose applications.

5 Paper capacitors

Values Miniature wire ended: 100 pF to 0.1 µF at 200 V to 750 V. Standard wire ended: 220 pF to 1 µF at 200 V to 750 V. Cans: 1 µF to 20 µF at 400 V to 1200 V. Various types of paper capacitors are illustrated in Figure 3.8.

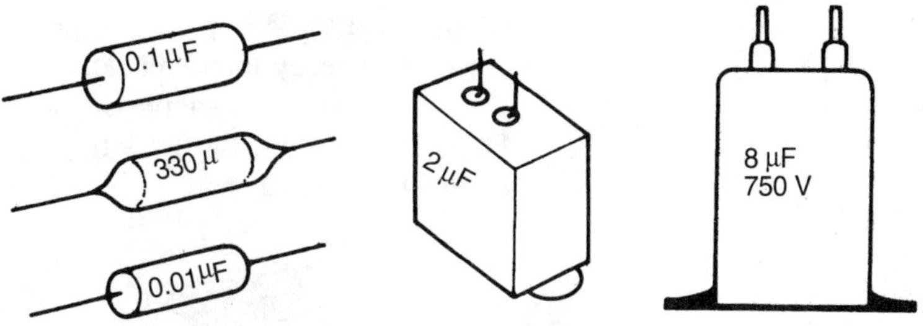

Figure 3.8 Various types of paper capacitors

Applications AC power circuits, DC blocking, motor phase shift, energy storage for strobe-flash units. General purpose AF work. Decoupling and smoothing at very high voltages and where ripple current may be high.

6 Polystyrene and silver mica

Values Polystyrene types: 10 pF to 10 000 pF at 30 V to 500 V. Silver mica types: 1 pF to 10 000 pF at 350 V. See Figure 3.9.

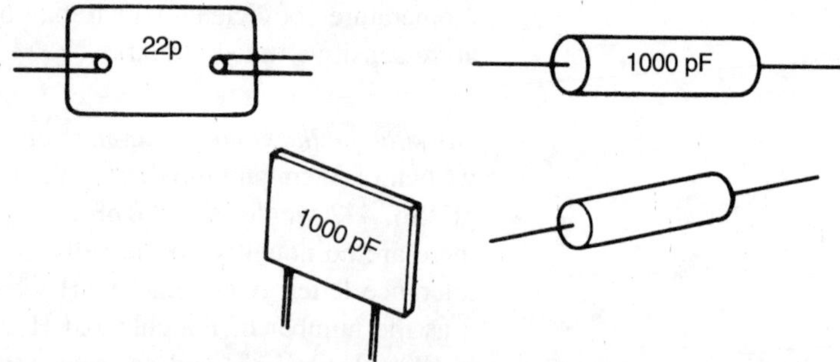

Figure 3.9 Left, polystyrene types; right, silver mica

Applications RF and AF filtering, tuning, AC coupling, RF decoupling. Generally used in high stability HF circuits. These types tend to be rather bulky for their capacitance but they have good, long-term stability and reasonably high working voltages. Like the other types, polystyrene and silver mica capacitors will work either way round in a circuit, but some types are marked to indicate which terminal should be connected to 0 V for screening purposes in HF circuits.

Other types

Bi-polar electrolytics are also available. These are non-polarised electrolytics specially designed for use in frequency dividing networks and other applications. Feed through capacitors are frequently used in HF circuits and are ideally suited to RF decoupling. Feedthrough capacitors can be used to decouple a power supply unit which is sensitive to RFI (Radio Frequency Interference).

Finally, variable and preset variables (trimmers) are also available. Figure 3.10 shows, on the left, a typical variable capacitor. A trimmer is shown on the right.

Figure 3.10 On the left, a typical variable capacitor. A trimmer is shown on the right

II – Calculations on capacitance

In the last section, we noted that the capacitance of a capacitor depends only upon its physical construction and that the terms are related in the following way:

$$C = \frac{\varepsilon_0 \varepsilon_r A}{d} \text{ farads}$$

Example 3.1

The parallel plates of an air-filled capacitor are everywhere 1mm apart. What must the plate area be if the capacitance is to be 1 F?

Solution

$$C = \frac{\varepsilon_0 \varepsilon_r A}{d} \text{ farads}$$

hence: $$A = \frac{C \times d}{\varepsilon_0 \varepsilon_r} \text{ m}^2$$

$$A = \frac{1 \times .001}{8.85 \times 10^{-12}} = 1.1 \times 10^8 \text{ m}^2$$

$1.1 \times 10^8 \text{ m}^2$ is equivalent to the area of a square sheet more than 6 miles on edge! This indicates that the farad is indeed a very large unit. In practical capacitors, however, the distance between the plates can be reduced by a very large factor and the dielectric can be increased at least a hundred times.

Voltage rating

We will see later that the amount of charge which can be held in a capacitor will depend on the value of the capacitor and the potential difference placed across it. However, we need also to consider the point at which the capacitor will break down; the closer the plates are together, the lower the breakdown voltage. Hence, it is important to specify not only the capacitor value, but its working voltage as well.

One way to increase the working voltage is to increase the distance between the plates. If this value is increased, however, the value of capacitance will be reduced. To compensate for this, the plate area must also be increased resulting in a physically larger capacitor.

For electrolytics, it is extremely important to choose an adequate voltage rating because these components can and do explode! It is also important to ensure that electrolytics are inserted into a circuit the correct way round, that they are not exposed to alternating current and that they are adequately rated for any ripple current which may exist.

Capacitors in series

It is possible to put capacitors in a circuit in series in order to increase the overall working voltage. When this is done, a potential divider should be

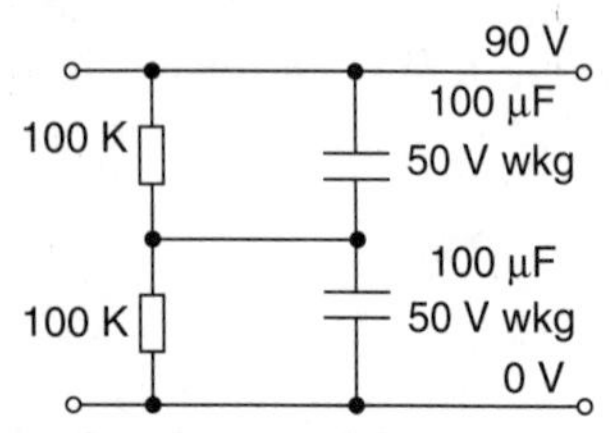

Figure 3.11 Capacitors in series

used to ensure that the working voltage of each capacitor is not exceeded. The circuit shown in Figure 3.11 shows a potential divider which may be used when capacitors are placed in series in order to increase the working voltage. The configuration will also result in reducing the value of *C* to 50 µF and this is explained later.

Capacitor summary chart

Type	*Non-polarised*			*Polarised*	
Property	Polyester	Mica	Ceramic	Aluminium	Tantalum
Values	0.01–10 µF	1 pF–0.01 µF	10 pF–1 µF	1–100 000 µF	0.1–100 µF
Tolerance	±20%	±1%	−25 to +50%	−10 to +50%	±20%
Leakage	small	small	small	large	small
Use	general	high frequency	decoupling	low frequency	low voltage

Ceramic/paper capacitors

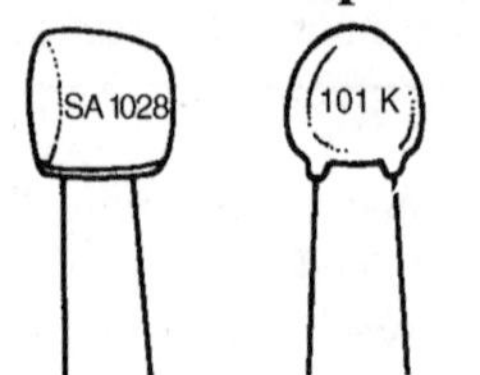

Figure 3.12 Ceramics

Many ceramic capacitors have a value coding which can only be described as unusual to say the least. The value is normally contained in three digits and is in pF. The problem is that the active three digits are often 'buried' inside manufacturers' serial numbers and a tolerance letter. In most cases, once the 3 digits have been identified, the first two constitute real numbers and the third is a multiplier – a number of zeros – as in the resistor colour code, e.g.

471 means 47 followed by one zero, i.e. 470 pF

Rather confusingly, therefore, 470 means 47 followed by no zeros, i.e. 47 pF. Some capacitors (often coloured green) may have a code like this: 2A102K. Ignore the first two alphanumerics, and note that the last letter (K) indicates the tolerance, hence:

2A | 102 | K = 1000 pF (= 0.001 µF or 1 nF)

2A 222 K = 2200 pF = 0.0022 µF

2A 472 K = 4700 pF = 0.0047 µF

2A 223 K = 22 000 pF = 0.022 µF

These capacitors are pictured in Figure 3.12.

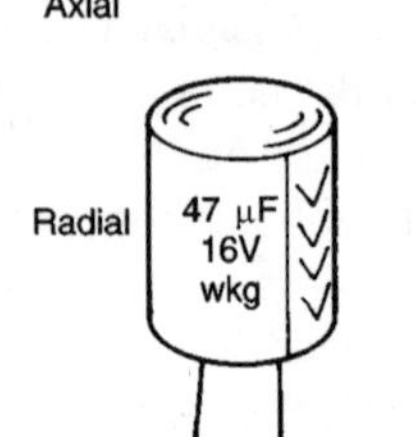

Figure 3.13 Axial and radial types

Electrolytics

Fortunately, most electrolytics have their values clearly marked, usually in µF; the working voltage and polarity is also generally indicated quite clearly. There are two basic designs for electrolytics, axial and radial, as shown in Figure 3.13.

The two types of package are electrically identical. The choice of which to use depends almost entirely on the circuit in which it is connected.

III – Voltage rating and $C = Q/V$

In the previous section we considered the importance of the voltage rating of capacitors. If too much voltage is placed across a capacitor then there is always a risk of the component breaking down. In the case of electrolytics, there is the additional risk of an explosion. In order to avoid this, capacitors are sold not only in terms of their capacitive value but in terms of their maximum voltage rating as well.

To understand what is happening, consider a large black plastic bin liner, for example. If you have a pile of screwed up newspapers, you can fill a typical bin liner with them. If you've still got some left afterwards, you can push on the top of the papers to compact them and produce more space for the other papers. If you continue to do this, pressing down and pushing more papers in, a point will come when the bag bursts and you have to start over again.

The same thing happens with capacitors except that this time you will be packing charge into the capacitor, the method of 'compacting' being to increase the potential difference across the plates. Attempt to 'compact' the electrons even more and the capacitor, like the bin liner, may explode.

The ratio between the charge stored in a capacitor and the potential difference across its plates is the value of capacitance. For a fixed value of capacitance, as the p.d. across it increases the amount of charge which can be stored in it also increases. The equation governing this is:

$$C = \frac{Q}{V}$$

Example 3.2

Calculate the value of a capacitor which holds 5×10^{-4} C of charge when 100 V p.d. exists across its plates.

Solution:

$$C = Q/V = 5 \times 10^{-4}/100 = 5\,\mu\text{F}$$

It has frequently been suggested that according to the equation $C = Q/V$ the value of a capacitor depends upon the voltage put across it. This is not true; capacitance depends only upon its physical dimensions – plate area and distance between plates, etc. From the equation, the unit of the farad is equal to so many coulombs per volt and this ratio is constant for a given capacitance. Hence, if, in the last example, the p.d. were raised to 200 V then the charge would be:

$$C = Q/V \quad \text{so} \quad Q = CV = 5 \times 10^{-6} \times 200 = 10^{-3}\text{C}$$

i.e. if the p.d. is doubled, the charge stored is doubled (in this case from 5×10^{-4} C to 10^{-3} C) the value of capacitance remaining the same.

Example 3.3

How much charge is stored in a 10 μF capacitor when a p.d. of 100 V is connected across its plates?

Solution

$$C = Q/V \text{ so } Q = CV = 10 \times 10^{-6} \times 100$$
$$= 10^{-3}\,\text{C}$$

Example 3.4

What is the p.d. across the plates of a 1500 μF capacitor if it stores a charge of 0.024 C?

Solution

$$C = Q/V \text{ so } V = Q/C = 0.024/1500 \times 10^{-6}$$
$$= 16\,V$$

Example 3.5

A polycarbonate capacitor of value 0.47 μF is required to store a charge of 1.5 × 10⁻⁴ C. What working voltage capacitor would you choose?

Solution

$$C = Q/V \text{ so } V = Q/C$$
$$V = 1.5 \times 10^{-4}/0.47 \times 10^{-6}$$
$$\text{and } V = 320\,V \text{ (approx.)}$$

IV – Capacitors in series and parallel

Remembering that resistors in series are simply added together and resistors in parallel have their reciprocals added together to find the reciprocal of the total resistance, a simple rule for capacitors is that 'it's the other way round'. In other words, for capacitors in parallel, the values are simply added together. When capacitors are connected in series, the reciprocal of the total value is equal to the sum of the reciprocals of the individual capacitors.

Example 3.6

Calculate the total capacitance of a 2.2 μF, 4.7 μF and 10 μF capacitor in parallel.

Solution

$$C_{tot} = C_1 + C_2 + C_3$$

so: $$C_{tot} = 2.2\,\mu F + 4.7\,\mu F + 10\,\mu F$$

$$= 16.9\,\mu F$$

Example 3.7

Calculate the value of the above capacitors in series.

Solution

$$\frac{1}{C_{tot}} = \frac{1}{C_1} + \frac{1}{C_2} + \frac{1}{C_3}$$

so: $$\frac{1}{C_{tot}} = \left[\frac{1}{2.2} + \frac{1}{4.7} + \frac{1}{10}\right] \times 10^6\,F$$

and: $$C = 1.3 \times 10^{-6}\,F = 1.3\,\mu F$$

Example 3.8

Calculate the total effective capacitance of the circuit shown in Figure 3.14.

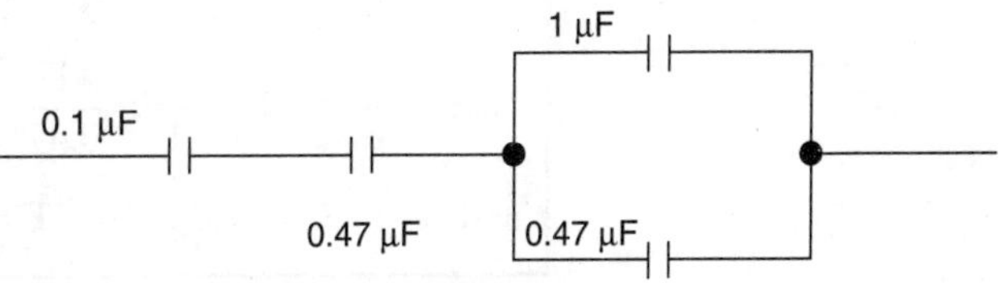

Figure 3.14 Circuit for Example 3.8

Try to work Example 3.8 out for yourself before looking at the answer which is given later. In the meantime let us discuss the maths used in Example 3.7. We've bracketed together the reciprocals of each capacitor value, then at the end, outside the brackets, we've multiplied everything by 10^6. This is because each individual capacitor has a value in µF (microfarads or a millionth of a farad) so it's convenient to take this common denominator outside of the brackets. The value should be $1/10^{-6}$ but since this is 10^6 it's even easier if we use that instead. However, all modern calculators have a button for reciprocals usually marked '1/x'. If you enter a number into the calculator, for example 2, then pressing the reciprocal button will return a value of 1/2 – a half or 0.5. Try it. If, then, you put in the first value in Example 3.7, 2.2 and then press the reciprocal button, the value you want is returned. A useful point here is to put that value in the calculator memory and then work out the next value in the same way, in this case 4.7. When that value is returned simply add it to what is already in the memory and then put the new total into the memory. After calculating and adding the third value, multiply the whole by 10^6.

What you have now is the reciprocal of the total, which is $1/C_{tot}$ so pressing the 1/x button one final time should give you the correct answer. What you should get is 1.3×10^{-6} F or 1.3 μF. If you didn't get the right answer, work through the instructions carefully and try it again. Once you've mastered this use of the calculator you can solve any series/parallel capacitor problem with confidence.

Solution (Example 3.8)

The two parallel capacitors simply have their values added together:

$$C_{tot} = 1\,\mu\text{F} + 0.47\,\mu\text{F} = 1.47\,\mu\text{F}$$

We now have three capacitors in series, the 0.1 μF, the 0.47 μF and the parallel pair which comes to 1.47 μF.

Solution

$$\frac{1}{C_{tot}} = \frac{1}{C_1} + \frac{1}{C_2} + \frac{1}{C_3}$$

so: $$\frac{1}{C_{tot}} = \left[\frac{1}{0.1} + \frac{1}{0.47} + \frac{1}{1.47}\right] \times 10^6\,\text{F}$$

and: $$C = 7.8 \times 10^{-8}\,\text{F} = 0.078\,\mu\text{F}$$

Example 3.9

How many 0.1 μF capacitors in series are required to make a 1000 pF capacitor?

Solution

$$1000\,\text{pF} = 1\,\text{nF} = 10^{-9} \quad \text{and:} \quad 0.1\,\mu\text{F} = 10^{-7}\,\text{F}$$

so: $$\frac{1}{10^{-9}} = \frac{1}{10^{-7}} \times X$$

and: $$X = 100$$

V – Charge and p.d. in series/parallel networks

Parallel network

If two capacitors are connected in parallel the p.d. across each must be exactly the same. Consider 8 μF and 16 μF capacitors in parallel.

If the p.d. is 100 V then for the 8 μF capacitor:

$$Q = CV = 8 \times 10^{-6} \times 100 = 8 \times 10^{-4}\,\text{C}$$

For the 16 μF capacitor:

$$\begin{aligned} Q &= CV \\ &= 16 \times 10^{-6} \times 100 \\ &= 1.6 \times 10^{-3}\,\text{C} \end{aligned}$$

Note For capacitors in parallel:

1 The potential difference across each is the same.
2 More charge is stored in the larger capacitor.

Series network

We now examine what happens when the same two capacitors are connected in series. Assume that the p.d. is 100 V across both of them. In this case, first work out the total effective capacitance and then obtain the charge from $C = Q/V$.

$$\frac{1}{C_{\text{tot}}} = \frac{1}{8\,\mu\text{F}} + \frac{1}{16\,\mu\text{F}}$$

Hence: $C_{\text{tot}} = 5.33 \times 10^{-6}\,\text{F}$

$C = Q/V$ so: $Q = CV = 5.33 \times 10^{-6} \times 100$

$$= 5.33 \times 10^{-4}\,\text{C}$$

This is the total charge in the circuit; it is also the charge on each capacitor. In this example, both the 8 μF and 16 μF capacitors have a charge of 5.33×10^{-4} C associated with them. This is often regarded as a surprising result, so consider the following:

1 Charge in a capacitive circuit is comparable to current in a resistive circuit. Each series component in a resistive circuit has the same current flowing through it. In series capacitive circuits, the same charge resides on each capacitor.
2 In our example, the innermost plates of each capacitor are connected together; they *MUST* therefore have the same charge, otherwise current would flow until it was. The charge on one plate is equal and opposite to the charge on the other plate so each capacitor has the same charge.

If the p.d. across both capacitors in series is 100 V then the p.d. across each may be calculated from $V = Q/C$.

For the 8 μF capacitor:

$$V = Q/C$$

so: $$V = \frac{5.33 \times 10^{-4}}{8 \times 10^{-6}}$$

and: $$V = 66.6\,\text{V}$$

For the 16 μF capacitor:

$$V = Q/C$$

so: $$V = \frac{5.33 \times 10^{-4}}{16 \times 10^{-6}}$$

and: $$V = 33.3\,\text{V (approx.)}$$

In calculations of this sort it is common to find that the sum of the p.d.s across each capacitor does not add up exactly to the total p.d. This is quite simply a matter of the number of decimal places used and does affect the validity of the result.

Note For capacitors in series:

1 The charge stored in each capacitor is the same (no matter what value capacitance they are).
2 The greater p.d. appears across the smaller capacitor.
3 The ratio of the capacitors in series is proportional to the p.d.s across them.

Example 3.10

5 μF, 10 μF and 20 μF capacitors are connected in series. With a p.d. of 100 V across all three, calculate:

(a) the charge stored in each capacitor
(b) the p.d. across each capacitor.

Solution

First, calculate C_{tot} and hence Q.

$$\frac{1}{C_{tot}} = \frac{1}{5\,\mu\text{F}} + \frac{1}{10\,\mu\text{F}} + \frac{1}{20\,\mu\text{F}}$$

$$= \left[\frac{1}{5} + \frac{1}{10} + \frac{1}{20}\right] \times 10^6$$

Hence: $$C_{tot} = 2.86\,\mu\text{F}$$

$$Q = CV = 2.86 \times 10^{-4}\,\text{C}$$

Hence:

(a) The charge in each capacitor is 2.86×10^{-4} C
(b) For the 5 μF, $V = Q/C = 57.2$ V
For the 10 μF, $V = Q/C = 28.6$ V
For the 20 μF, $V = Q/C = 14.3$ V

Check: $57.2 + 28.6 + 14.3 = 100$ V (approx.)

VI – Charge and discharge curves

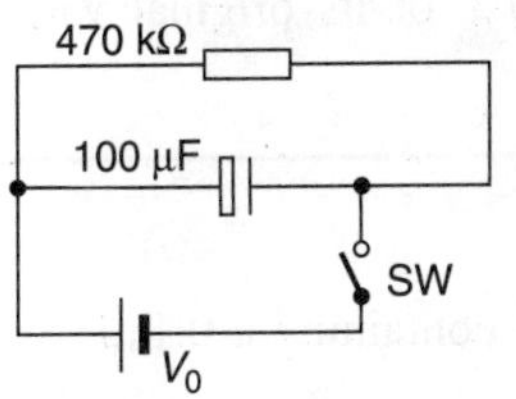

Figure 3.15 Circuit for studying the rate of discharge of a capacitor

When a potential difference is applied across the plates of a capacitor, charge flows into it. However, it is not a uniform flow. The charge flows in very quickly at first, but then gradually slows down, flowing at its slowest as it reaches the maximum value of charge it can hold. This non-uniform rate of flow of charge is known as exponential charging.

The same thing happens when a capacitor is discharged. The flow of charge is rapid at first but it gradually slows down when the discharge approaches completion. This is called the exponential discharge of a capacitor. A typical circuit for studying the rate of discharge of a capacitor is shown in Figure 3.15.

The switch, SW, is first closed for a short period of time until the capacitor is fully charged. The switch is then opened and the capacitor discharges through the resistor. A high impedance digital voltmeter (DVM) placed across the capacitor will then show the exponential reducing voltage.

In this circuit, the product *RC* is known as the 'time constant'. The rate of charge or discharge in a capacitance resistance circuit is determined by the time constant. Since the decay is exponential, the total time taken to discharge the capacitor completely cannot be computed. This is because, after a time equal to the time constant, only a proportion of the charge will have decayed, so in theory the capacitor never discharges completely at all. To get round this problem, a time in seconds equal to the time constant is used.

The time constant is represented by the Greek letter τ (tau). Using the component values given in Figure 3.15, the time constant is calculated as follows:

$$\begin{aligned}\tau &= R \times C\\ &= 100\,\mu\text{F} \times 470\,\text{k}\Omega\\ &= 100 \times 10^{-6} \times 470 \times 10^{3}\\ &= 4.7 \times 10^{-4} \times 10^{5}\end{aligned}$$

hence: $\tau = 47$ seconds

Note (incidentally) in the above example, on line four of the calculation we have separated out the numbers from the powers of ten. This makes the calculation so much easier – in fact, working like this, you don't even need a calculator.

The significance of the time constant is that a certain percentage of charge will always decay during that time, which in this case, is 47 s. If we had used a 68 μF capacitor with a 1MΩ resistor, then the time constant $\tau = 68 \times 10^{-6} \times 10^{6} = 68$ s. In that time, the same percentage of charge will have decayed. The value of that percentage is 37%. In other words, in any capacitor–resistor circuit of this type, after a time equal to the time constant has expired, the charge in the circuit will have reduced to 37% of its original value.

To show how this is derived requires a rather complicated equation. You will never need to use it, however, so it is sufficient simply to remember the all-important definition.

> When a fully charged capacitor is subsequently discharged, the voltage across its terminals will fall to 37% of its original value, in a time equal to the time constant.

Example 3.11

(a) Calculate the time constant of a circuit containing a 0.1 μF capacitor and 2.2 MΩ resistor.
(b) If the initial voltage across the capacitor is 10 V, what is its value one time constant after the switch is opened?

Solution

(a) $\tau = RC = 0.1\,\mu\text{F} \times 2.2\,\text{M}\Omega$

$$= 0.1 \times 10^{-6} \times 2.2 \times 10^{6}$$

$$= 2.2 \times 10^{-7} \times 10^{6}$$

so: $\tau = 0.22$ s

(b) $V = V_o \times 0.37 = 0.37 \times 10 = 3.7$ V

Example 3.12

(a) Calculate the time constant of a circuit containing a 100 μF capacitor and 500 kΩ resistor.
(b) If the battery voltage is 12 V calculate:
   (i) the voltage across the capacitor one time constant after the battery is disconnected
   (ii) the time after switch-off when this value is reached.

Solution

(a) $\tau = RC = 100 \times 10^{-6} \times 500 \times 10^{3} = 50$ s
(b) (i) $V = V_o \times 0.37 = 12 \times 0.37 = 4.44$ V
(ii) One time constant, 50 s.

In the case of the exponential charge of a capacitor, the solution shows that the voltage across a capacitor as it charges will rise to 63% of its final value in a time equal to the time constant.

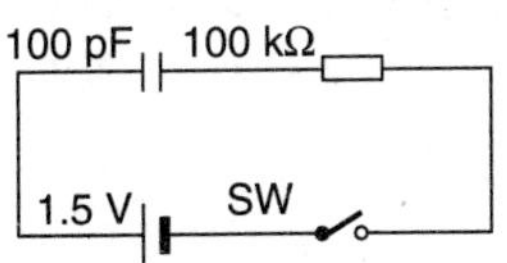

Figure 3.16 Circuit for studying the rate of discharge of a capacitor

> Example 3.13
>
> Consider the circuit shown in Figure 3.16.
>
> (a) Calculate the time constant of a circuit containing a 100 pF capacitor and 100 kΩ resistor.
> (b) If the supply voltage is 1.5 V calculate the p.d. across the capacitor after one time constant.
>
> *Solution*
>
> (a) $\tau = RC = 100 \times 10^{-12} \times 100 \times 10^{3} = 10^{-5}$ s
> (b) $V = V_o \times 0.63 = 1.5 \times 0.63 = 0.945$ V

Both the charge and discharge of a capacitor may be shown graphically as in Figure 3.17.

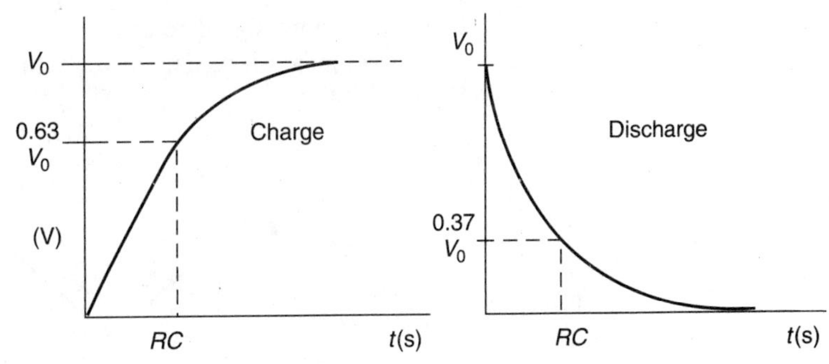

Figure 3.17 Charge and discharge curves for a typical *RC* circuit

Figure 3.18 An integrator

In the figure, the exponential nature of the charge and discharge of a capacitor is clearly shown. Additionally, you will also see marked the voltage across the capacitor after one time constant has elapsed in each case.

VII – Integrators and differentiators

The integrator

In the simple circuit of Figure 3.18, a capacitor C can be charged via the resistance R, by the battery V. From the discussion on capacitor charge and discharge patterns in the last section, we can (to some extent) predict what the output waveform will be. As the capacitor charges it will follow the same exponential rise we've already seen, and reach 63% of its final value in the time, τ, equal to the time constant.

If the values of R and C are given then we could draw a sketch graph of the rising output which occurs after the switch is closed. If the time constant is large (say about 10 s) then the capacitor will be fully charged after 20 s and no more current flows in the circuit. If the switch is then opened, the capacitor will discharge following the familiar exponential decay we have noted previously. If the same pattern is followed repeatedly, then the output will look something like that shown in Figure 3.19.

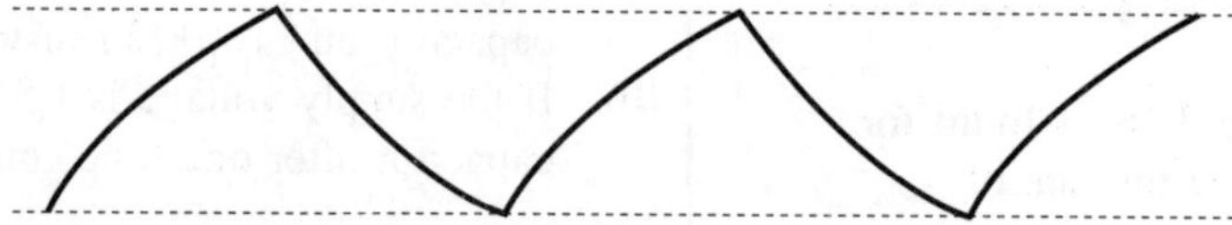

Figure 3.19 Output waveform from an integrator

Taking this one step further, if, instead of connecting a battery and a switch to the input, a square wave is applied (provided it is of the appropriate frequency) then a continuous waveform like the one shown in Figure 3.19 will appear at the output.

If the input frequency is increased (making the period smaller) the capacitor won't have time to approach even 63% of its final value before the input signal returns to zero. In this case, the output waveform will become a triangular wave with almost straight sides as shown in Figure 3.20.

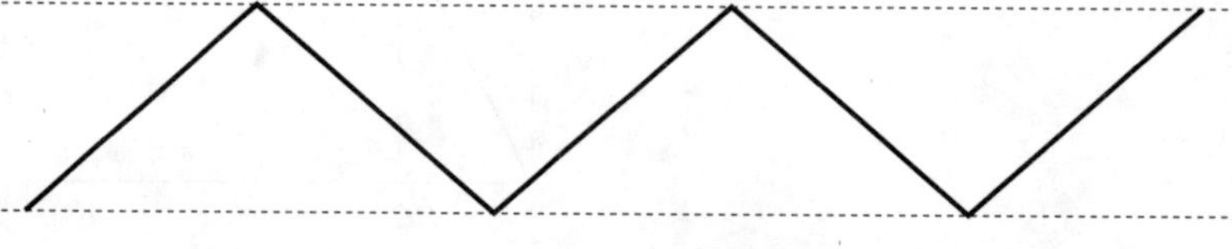

Figure 3.20 Output waveform from an integrator when the product RC is much greater than the period of the input waveform

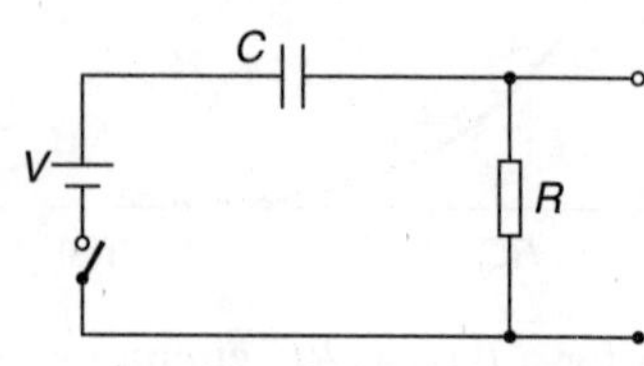

Figure 3.21 A differentiator

The differentiator

If the resistor and capacitor change places, then the circuit becomes a differentiator as shown in Figure 3.21.

Again, the precise shape of the output waveform depends upon the comparative values of RC and the period of the applied waveform. However, for given values of RC, if the input frequency is very high, then the output will approach a square wave. If the input frequency is very low then the output will be a series of pulses, as shown in Figure 3.22.

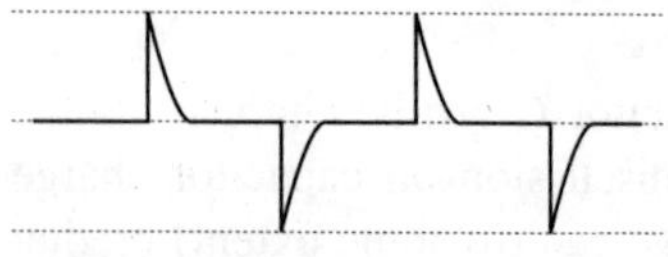

Figure 3.22 Typical output waveform from a differentiator

In summary, for a given square wave input, an integrator produces an output waveform which is triangular, and a differentiator produces a series of positive- and negative-going pulses.

Questions

1 Calculate the value of a capacitor whose plate area is $2.5\,m^2$ and distance between plates is 0.1 mm. Assume $\varepsilon_r = 1$.

2 A parallel plate capacitor has a common area of $100\,cm^2$ and a plate separation of 2 mm. Calculate the capacitance if $\varepsilon_r = 1$.

3 A 0.01 μF paper capacitor has a common area of $750\,cm^2$. Find the thickness of the paper if the dielectric constant (ε_r) of the paper is 2.

4 Calculate the value of a mica capacitor whose plates have an area of $40\,cm^2$ and are separated by 0.1 mm. Assume $\varepsilon_r = 5$.

5 Calculate the value of a polystyrene capacitor, if the plates are $10\,cm^2$ in area, separated by a distance of 0.5 mm and the value of $\varepsilon_r = 2.55$.

6 What type of capacitor would most commonly be used in a power supply unit for smoothing?

7 (a) State four important precautions to be taken when using electrolytic capacitors.
(b) For which two types of capacitor would you expect to find the voltage rating indicated by a colour code?
(c) Which type of capacitor in common use has the highest voltage rating?

8 What is the capacitance of a paper capacitor which stores a charge of 0.01 C at 1200 V?

9 How much charge flows into a tantalum capacitor of value 2.2 μF if a p.d. of 15 V is applied across it?

10 A polycarbonate capacitor of value 0.47 μF is required to store a charge of 1.5×10^{-4} C. What working voltage capacitor would you choose?

11 Which capacitor can safely hold more charge?

(a) a 10 000 pF, 250 V working, polystyrene capacitor
(b) a 0.47 μF, 100 V working ceramic capacitor.

12 A tantalum capacitor stores 5×10^{-4} C of charge at 10 V. What is its capacitance?

13 Calculate the value of the capacitor which could replace a 10 μF and 15 μF capacitor in series.

14 Calculate the value of the single capacitor which could replace three 22 μF capacitors in series.

15 What value capacitor in series with a 0.1 μF capacitor would produce a 0.032 μ capacitor?

16 What is: (a) 100 nF in μF, (b) 100 pF in nF, (c) 10 000 pF in μF, (d) 1000 μF in F?

17 How many 1 μF capacitors would need to be connected in parallel in order to store a charge of 1 C with a p.d. of 300 V across the capacitors?

18 A 6 μF capacitor is connected in series with a 4 μF capacitor and a potential difference of 200 V is applied across the pair.

(a) What is the charge on each capacitor?
(b) What is the p.d. across each capacitor?

The same two capacitors are now connected in parallel.

(c) What is the charge on each capacitor?
(d) What is the p.d. across each capacitor?

19 A parallel plate capacitor has circular plates of radius 8 cm which are air-spaced to a distance of 1 mm. Assuming that $\varepsilon_0 = 8.85 \times 10^{-12}\,\text{Fm}^{-1}$, what charge will appear on the plates if a potential difference of 100 V is applied?

20 Three capacitors of value 10 μF, 20 μF and 70 μF are connected in parallel. If a p.d. of 200 V is applied to this parallel combination, how much charge is stored in each capacitor?

Multiple choice questions

1 A capacitor is marked with a value of 10 μF. It is *most* likely to be

A paper
B polyester
C polystyrene
D electrolytic

2 A 100 nF capacitor is equivalent to

A 1 μF
B 0.1 μF
C 0.01 μF
D 0.001 μF

3 The substance separating the plates of a capacitor is called the

A dielectric
B diethylide
C electrolytic
D electrolyte

4 A 'trimmer' is a

A variable capacitor
B preset variable capacitor

C miniature electrolytic
D miniature tantalum

5 A 0.01 μF capacitor is equivalent to

A 1 nF
B 10 nF
C 100 nF
D 1000 nF

6 The value of a capacitor can be increased by

A increasing the plate separation
B decreasing the plate separation
C making the plates thicker
D making the plates thinner

7 One example of a polarised capacitor is

A electrolytic
B paper
C polycarbonate
D ceramic

8 A capacitor is marked with '2A 472K'; its value is most likely to be

A 472 pF
B 472 nF
C 4700 pF
D 4700 nF

9 A capacitor is colour coded and has no printed value on it. It is most likely to be

A polycarbonate
B polyester
C polypropylene
D polystyrene

10 The main advantage of an electrolytic capacitor over non-electrolytic types is that it has a

A low insulation leakage
B high stability
C high capacitance for a small volume
D large DC working voltage

11 A capacitor marked 4.7 nF is equivalent to

A 4700 pF
B 47000 pF

C 0.047 μF
D 0.47 μF

12 Charge is measured in

A Volts
B Amps
C Joules
D Coulombs

13 100 V is placed across a 2 μF paper capacitor. The charge held will be

A 0.2 mC
B 2.0 mC
C 20 mC
D 200 mC

14 A capacitor coded 2A 471K is most likely to be

A 47 pF
B 470 pF
C 471 pF
D 4700 pF

15 A capacitor holds 0.02 mC of charge with 10 V across it, the capacitor value is

A 0.2 μF
B 2 μF
C 20 μF
D 22 μF

16 The breakdown voltage of a capacitor is related to

A plate area
B plate thickness
C plate separation
D plate resistance

17 In order to obtain a working voltage of 100 V, two 68 V working capacitors may be put

A in series
B in series with a potential divider in parallel
C in parallel
D in series parallel

18 An integrator contains a 1 MΩ resistor and a 10 nF capacitor. The time constant is

A 0.01 s
B 0.1 s

C 1.0 s
D 10 s

19 A capacitor has one terminal which is marked with a minus sign and an arrow. It is most likely to be

A a paper capacitor
B a ceramic capacitor
C a tantalum capacitor
D an electrolytic capacitor

20 A fully charged capacitor is placed across a 100 kΩ resistor and proceeds to discharge through it. After a time equal to the product *RC* of the components in the circuit, the voltage across the capacitor will have reduced to

A 33.3%
B 37%
C 50%
D 63%

4 Magnetism and electromagnetism

Simple magnetism

Almost everyone must be familiar with an ordinary bar magnet; it has two poles called NORTH and SOUTH where all the magnetism seems to be concentrated.

The magnets we normally see and use are made artificially. There are some natural magnets to be found in certain parts of Asia Minor, which are rough pieces of a rock-like substance containing an iron ore. The ancients called this rock-like substance 'lodestone' or lead-stone because they noticed that if it were freely suspended by a cord, it always pointed in one direction. This was a very useful 'lead' or guide when travelling in desert areas. The lodestone was also used by sailors for navigation purposes. It was found that a piece of iron could be made into an artificial magnet by stroking it with a lodestone. The north pole of a magnet is actually the 'north-seeking' pole, hence, the Earth acts as if it had a huge bar magnet inside it, with its south pole in the northern hemisphere. The field around a bar magnet is non-uniform but the Earth's magnetic field is considered to be uniform.

Lodestone is a certain form of iron ore, also known as MAGNETITE after the place in which it was discovered around 600BC. It is said to be magnetic because it attracts some other metals, notably iron, cobalt and nickel. Steel can be magnetised because it contains iron. Many magnets are made from ALNICO which is a special alloy of aluminium, nickel and cobalt. It is used to make very strong, permanent magnets. Listed below is a summary of the important facts about magnets.

1 A magnet is a metal which attracts *some* other metals.
2 A magnet does not attract all substances, only a few.
3 The most common magnetic substances are IRON, COBALT and NICKEL. Steel is magnetic because it contains iron.
4 Magnetism will act through materials, e.g. paper, glass.
5 The poles of a magnet are the strong parts at the ends. The two poles are called the NORTH and SOUTH (seeking).
6 Like poles repel; opposite poles attract.
7 A magnet can be made by stroking a magnetic material with another magnet or by placing it in a DC coil.
8 A magnet may be destroyed by heating, strong mechanical vibrations or by being placed in an AC coil.
9 Some uses of magnets: compasses, magneto (electricity generator), some types of microphones, loudspeakers, fridge doors, moving iron and moving coil gramophone pickups, some electric motors;

in televisions: ion traps, pin-cushion correction, linearity coils, convergence coils; toys; automatic can openers and even paper clip dispensers!

Magnetic field patterns

Any conductor which has a current flowing through it has a magnetic field around it. This can be demonstrated by the use of plotting compasses which have a transparent base. When placed in a magnetic field, the compass aligns itself along a magnetic line of force with its north pole pointing towards the south pole of the field.

By moving the compass along the line of force and marking with a pencil the direction in which it points, the magnetic field can be traced out. A typical field pattern is shown in Figure 4.1 and is readily recognised by most people.

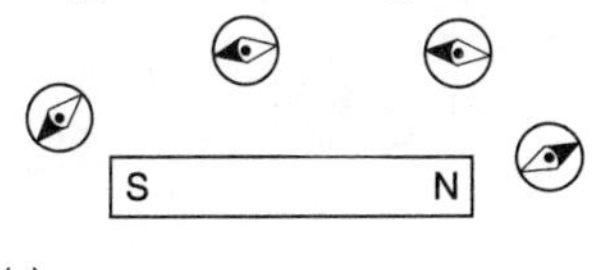

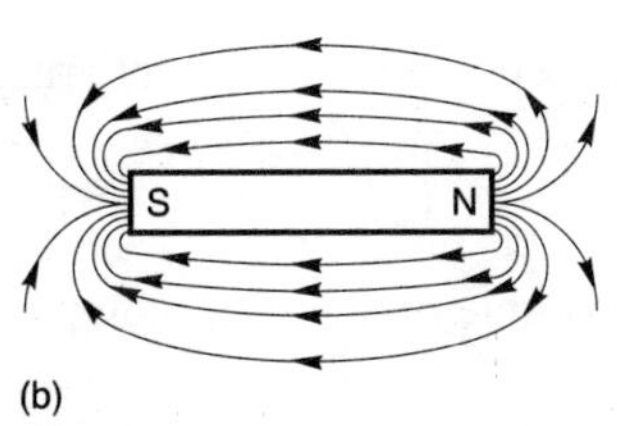

Figure 4.1 (a) Using plotting compasses to trace out the magnetic field of an ordinary bar magnet. (b) A typical bar magnet field pattern

If a stiff copper wire is placed vertically piercing a piece of card, such that the card acts like a 'stage' for the plotting compass, the magnetic field produced by a current flowing through the wire may be traced out. It will have the shape of a series of concentric circles as shown in Figure 4.2.

If the piece of wire is now bent into a ring as shown in Figure 4.3 the current flowing through it will produce a concentrated magnetic field flowing through its centre. The direction of the magnetic field depends upon the direction of current flow which can be worked out using Maxwell's corkscrew rule. The right hand is used to grasp the wire such that the thumb points in the direction of the current; the direction of the fingers then gives the direction of the field.

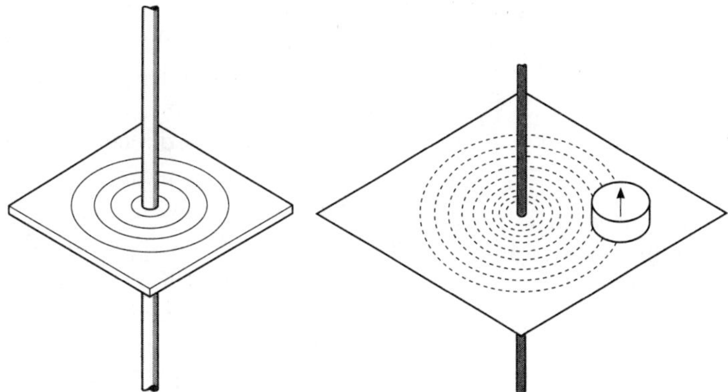

Figure 4.2 The magnetic field produced by a current-carrying conductor

If two rings are used, the magnetic field through the centre becomes even stronger. If many rings are placed side by side, by winding copper wire round something like the cardboard inside of a loo roll as shown in Figure 4.4, a 'solenoid' is produced. This is basically a current-carrying coil of wire which has a concentrated magnetic field inside it. The solenoid then produces a magnetic field pattern which is very similar to that

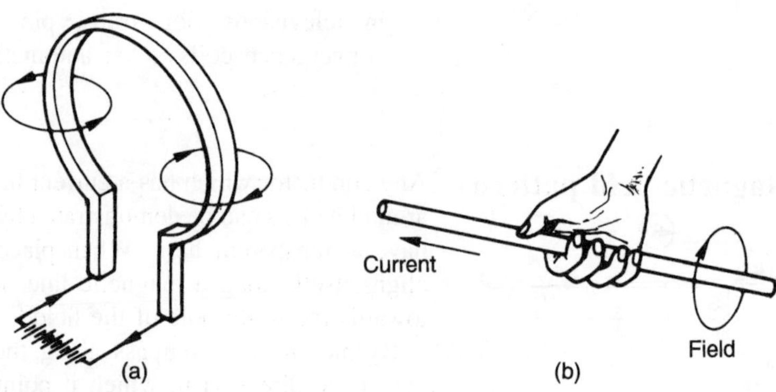

Figure 4.3 (a) The magnetic field associated with a current ring. (b) Maxwell's Corkscrew Rule

Figure 4.4 The solenoid consists of many turns of copper wire wound around a cylindrical former

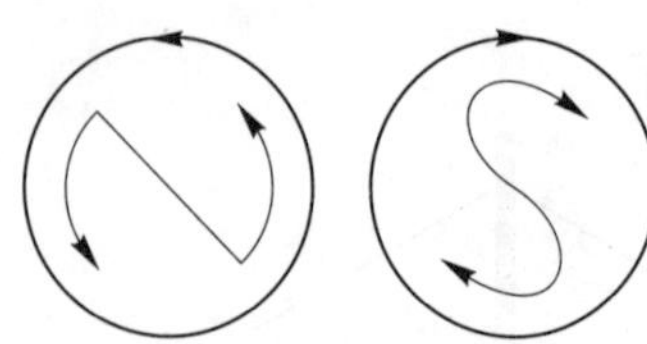

Figure 4.5 How to determine the poles of an electromagnet

produced by an ordinary bar magnet of comparable dimensions, complete with north and south poles.

If a circular cross-section bar of soft iron is now introduced down the centre of the solenoid, the magnetic field associated with it becomes even stronger and an electromagnet is produced. Such a bar of metal is frequently referred to as the 'core' in this context. When current flows through the coil, the iron bar becomes magnetised; when the current is switched off, the iron bar loses most of its magnetism. There is often some residual magnetism left in the core when the current has been switched off, depending on the precise nature of the metal introduced into the coil. If steel or alnico is placed inside the coil it becomes permanently magnetised and it is in this way that magnets may be manufactured.

In order to determine the poles of an electromagnet it is necessary to note the orientation of the windings (clockwise or anti-clockwise, looking at the end) and the direction of the current. Fortunately there is a very simple method of determining this as shown in Figure 4.5.

If the coil is wound in a clockwise manner when viewed from the end, and the current flows in the direction indicated by the arrows, then a south pole is produced. The exaggerated shape of the 'N' shows how a north pole is produced. The electromagnet can be made stronger by increasing the number of turns in the coil, in several layers if necessary, and by increasing the current flowing through it. A device such as this can be referred to as a coil, a solenoid or an electromagnet. They all mean much the same thing and which is used depends only on the application for which it is intended. Solenoids may be used to convert electrical energy into mechanical energy, in a relay for example. The relay consists of a solenoid and an armature with a set of contacts. Using such a device may enable a high current, mains electric lamp to be switched on and off by a 3 V battery. Provided the coil is adequately rated, and the contacts designed to handle the voltage and current, this could be achieved very simply.

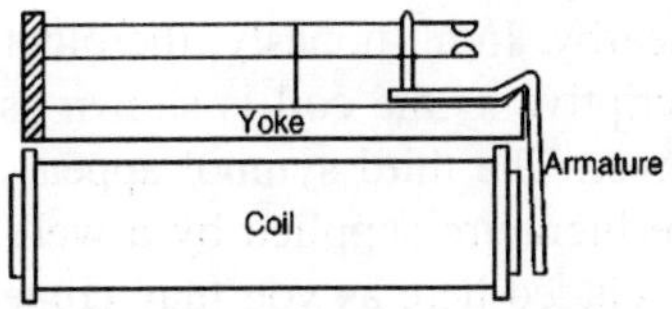

Figure 4.6 A typical 'Post Office' relay

Relays come in all shapes and sizes but perhaps one of the simplest representations is shown in Figure 4.6. This sort of relay was used extensively by the Post Office when they were in charge of telecommunications. Very few now remain since modern, electronic methods of switching which are faster and more reliable are used by British Telecom. However, many motor cars still use relays to supply the starter motor, a device which passes a huge current via the relay contacts, but which only requires the small coil current activated by the ignition key switch.

The central locking now employed by most new cars and other automatic door locks also uses the electromagnetic effect. It usually has a piston which moves in and out of the coil as it is activated.

Relay coils can be made to operate on a variety of currents and voltages but they tend to be designed to operate at low voltages and currents, while the contact switch operates on much higher ones. This provides a simple method of allowing a low current device – typically a small transistor – to switch a much higher current device. It is even possible to construct a simple, 'touch sensitive' circuit using a transistor and relay. The touch pad is simply two terminals, perhaps of copper laminate, separated by a small gap of, say, 3 or 4 mm. When a finger is placed across the gap, the small current flowing into the transistor is enough to turn it on and activate the relay.

The circuit is shown in Figure 4.7, but it might not make too much sense until you have studied the action of the transistor as a switch in Chapter 9, page 138, where all is revealed. Meanwhile, 'back emf', referred to later on in this chapter, may have a possibly harmful effect on the transistor, and it is for this reason that the diode is included across the relay coil.

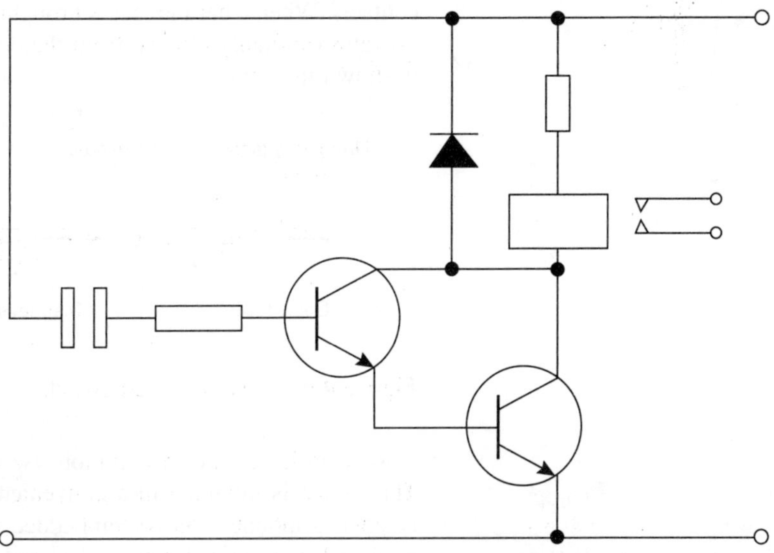

Figure 4.7 A touch sensitive circuit employing a relay

The standard relay symbol is shown in Figure 4.8a and is simply a rectangular box with the contacts shown nearby. In all honesty, the older symbol shown in Figure 4.8b is more descriptive as the coil is shown as it ought to be. It is rarely used now, however. The third symbol appears to be American in origin being used in the literature supplied by a well-known high street electronics outlet; it is included here as you may come across it sometimes. The contacts may be normally open (no), normally closed (nc) or change over (co) types and some of them are shown in the diagram.

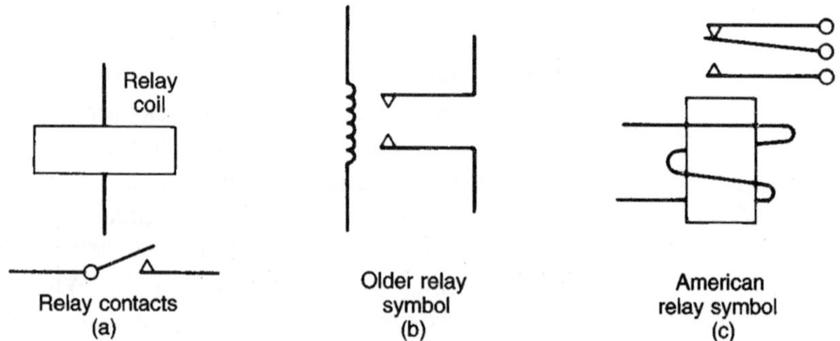

Figure 4.8 (a) Standard relay symbol. (b) Older relay symbol. (c) An American relay symbol

Reed switches

The reed switch consists of thin strips of easily magnetisable and demagnetisable materials, enclosed in a small, glass envelope. The glass tube usually contains an inert gas such as nitrogen to reduce corrosion of the contacts. When a bar magnet is brought close to the device, the reed switch operates (normally closing from the open position). A typical reed switch is shown in Figure 4.9.

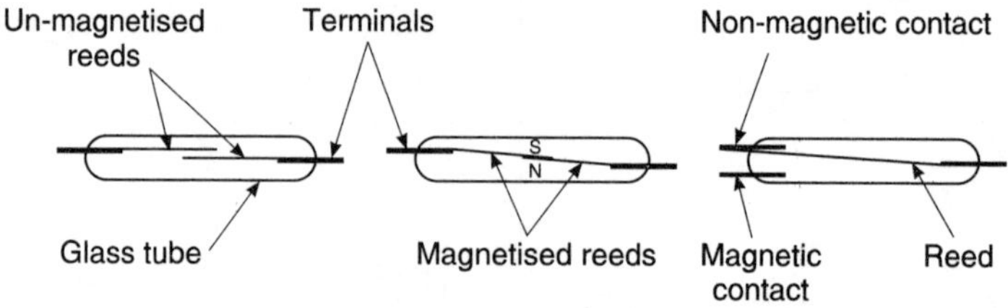

Figure 4.9 A typical reed switch

Reed switches are in common use as part of a burglar alarm system. The switch is mounted in a convenient position above a door, while the magnet is mounted on (or embedded in) the door itself. When the door is closed, the magnet causes the reed switch to close; when the door is opened, the reed switch opens and this sounds the alarm.

Other types of reed switch are designed to be used with a coil. The reed switch sits inside the coil such that, when the coil is energised, the reed switch operates. The reeds separate when the coil current is turned off. If an alternating current is supplied to the coil, the reed switch opens and closes at the frequency of the supply. In the changeover (co) type, the reed is attracted from the non-magnetic contact to the magnetic one.

Earth Leakage Circuit Breakers (ELCBs)

These are used as safety devices in mains electrical appliances. In one type, current passes to earth through a relay-type 'trip coil' when, for example, the metal case of the appliance becomes 'live' due to a fault. In the type illustrated in Figure 4.10, the fault current energises the coil and a rod inside it moves and opens a switch which can be set to break the circuit before the case rises to a lethal voltage.

Other similar devices (e.g. a residual current device or RCD) measure the current going into the appliance and the current coming out. Any discrepancy indicates a fault and trips the switch.

Force on a current-carrying conductor

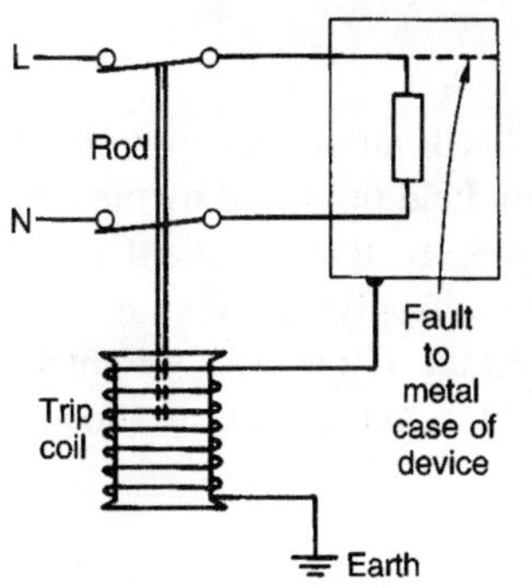

Figure 4.10 One kind of earth leakage circuit breaker (ELCB)

This effect can be shown quite simply by placing a wire inside the field of a very strong, horseshoe magnet. When current is passed through the wire, a magnetic field is created around it, as we have already seen. The interaction between this field and the field produced by the horseshoe magnet causes an imbalance and the force produced causes the wire to move. This important phenomenon forms the basis of operation of moving coil meters and electric motors. A typical apparatus for demonstrating the effect is shown in Figure 4.11.

The direction of the force and hence movement of the wire can be predicted by the use of Fleming's Left Hand Rule. If the thumb, and first and second fingers of the left hand are held mutually at right angles, then:

the **F**irst finger indicates the direction of the **F**ield,
the se**C**ond finger indicates the direction of the **C**urrent, and
the thu**M**b indicates the direction of **M**otion.

The direction of the magnetic field follows the usual convention of running from north to south pole. The rule is shown diagrammatically in Figure 4.12.

By experiment it can be shown that the force experienced by the current-carrying conductor depends upon three things:

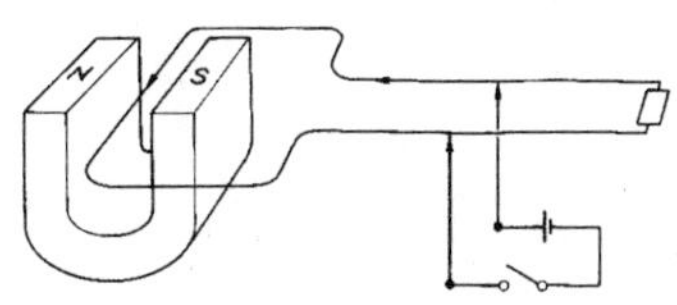

Figure 4.11 Demonstrating the force acting on a current-carrying conductor in a magnetic field

1. The current, I, which, if increased, increases the force.
2. The strength of the magnetic field, B, which, if increased, also increases the force.
3. The length of conductor, l, actually in the magnetic field, which, if increased, also increases the force.

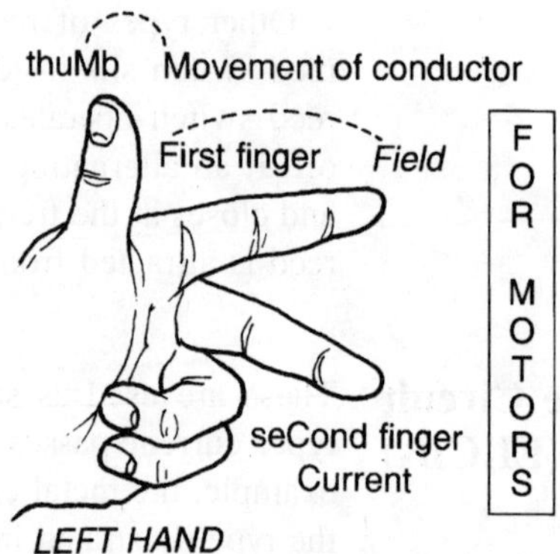

Figure 4.12 Fleming's Left Hand Rule

The current is measured in amps, the magnetic field in tesla, and the length of the conductor in metres. If these units are used then the force (in newtons) is simply the product of the three quantities involved.

$$F = B \times I \times l \quad \text{(newtons)}$$

The equation is true, however, only if the current-carrying conductor is at right angles to the magnetic field.

Parallel conductors

Two parallel wires carrying parallel currents will attract each other. This is because each conductor lies in the magnetic field produced by the other. Drawing in the magnetic lines of force, we see that the field can be regarded as acting perpendicular to the wire. Applying Fleming's Left Hand Rule shows that the two wires will attract. If the two currents are flowing in the opposite direction, the conductors repel. This is shown in Figure 4.13.

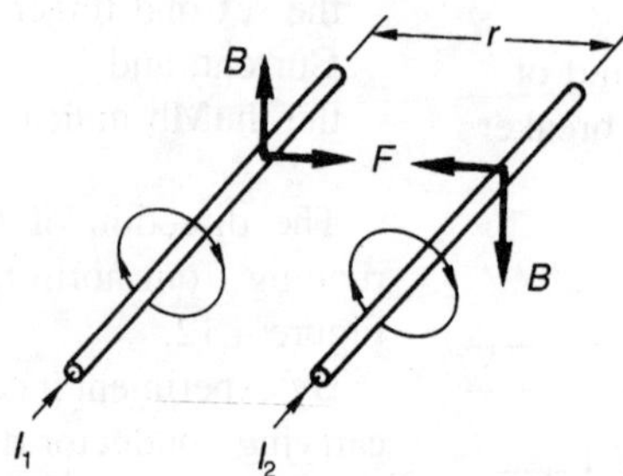

Figure 4.13 Parallel wires carrying parallel currents attract each other

The loudspeaker

This familiar device works on the principle of a current-carrying conductor in a magnetic field experiencing a force. The field is invariably produced

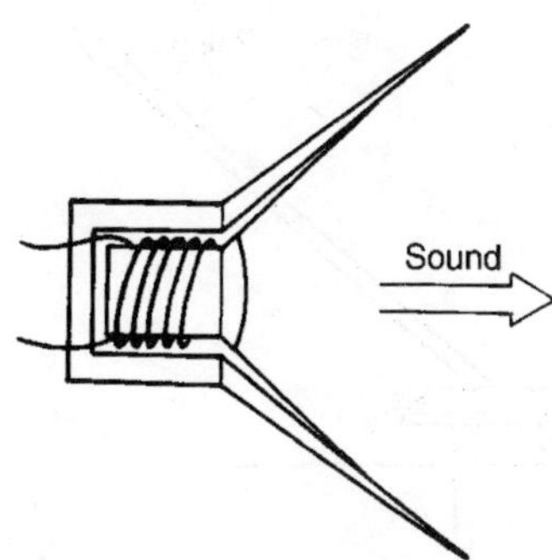

Figure 4.14 A loudspeaker

by a circular magnet; the current-carrying conductor (called a speech coil) sits around the magnet. So, when current flows through the speech coil, it experiences a force, pushing the cone attached to it outwards. The cone pushes the air around it creating a sound wave. This is shown in Figure 4.14.

If a loudspeaker is connected to a 4 V DC supply, the cone can be seen to move forwards or backwards, depending on which way round the supply is connected. When the loudspeaker cone moves forwards, the terminal connected to the positive of the supply is usually marked with a '+' or a red spot or similar marking. Hence, when several loudspeakers are connected together in the same enclosure, or used as stereo speakers, all the 'positive' terminals of the loudspeakers must be connected in the same sense. When this condition is satisfied, the loudspeakers are said to be in phase.

If the speakers are connected in parallel, all the '+' terminals are connected together; if they are in series, opposite terminals are connected together. This is shown in Figure 4.15.

Impedance

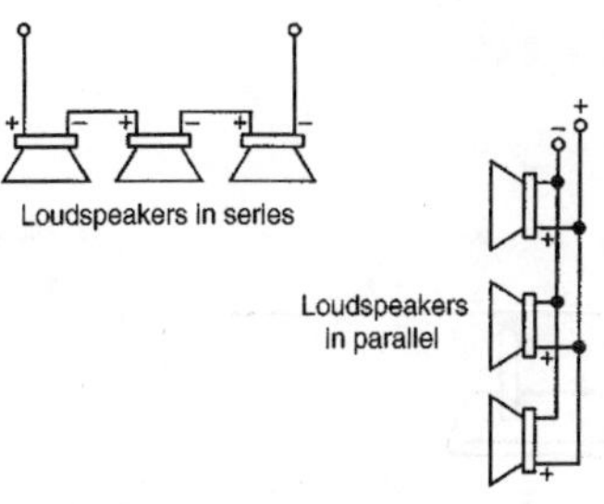

Figure 4.15 Loudspeakers in series and parallel

When a current flows through a coil, a 'back emf' is produced (it is explained how this arises later). The back emf may be regarded as a current in the coil which opposes the current being supplied to it. This opposition to a current has a similar effect to resistance – that of reducing the current in the circuit – but is produced in a quite different way. When an AC current is applied to a coil, a continuous back emf is produced. This 'AC resistance' is called reactance. Although it is usually very small, the coil also has some resistance, so both this and the reactance must be taken into consideration. The combined effect of the resistance and reactance of a coil is called impedance, its value is often marked on a loudspeaker, and is also measured in ohms. Note that when loudspeakers are connected in series or parallel, the effective impedance changes (like the resistance of resistors connected in series or parallel) and this must be taken into consideration so that the impedance of the loudspeaker array matches the output impedance of the amplifier to which it is connected.

Moving coil meters

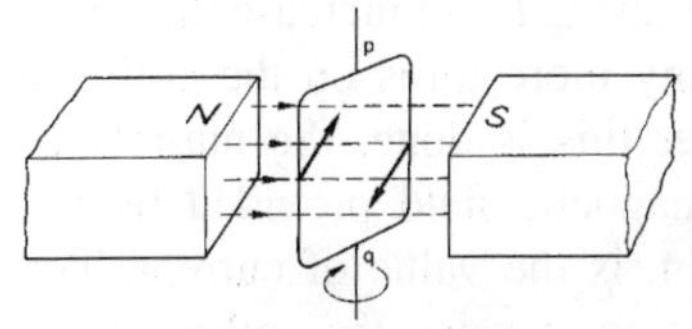

Figure 4.16 A current-carrying coil in a magnetic field

We have already seen how a current-carrying conductor in a magnetic field experiences a force ($F = BIl$) the direction of which is given by Fleming's Left Hand Rule. In Figure 4.16 a coil of just a few turns (for clarity) is placed in the magnetic field provided by two bar magnets.

The field flows from north to south and the current goes up on the left hand side and down on the other. From Fleming's rule, there will be a force acting inwards on the left hand side and a force outwards on the right hand side. If the wires leading into the coil are supported on pivots at p and q, the coil will turn in the magnetic field. Two forces acting in this way produce a 'couple' and the force producing this circular motion is called torque.

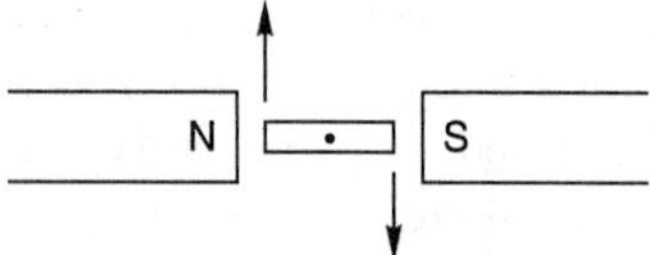

Figure 4.18 Forces acting on a current-carrying coil in a magnetic field

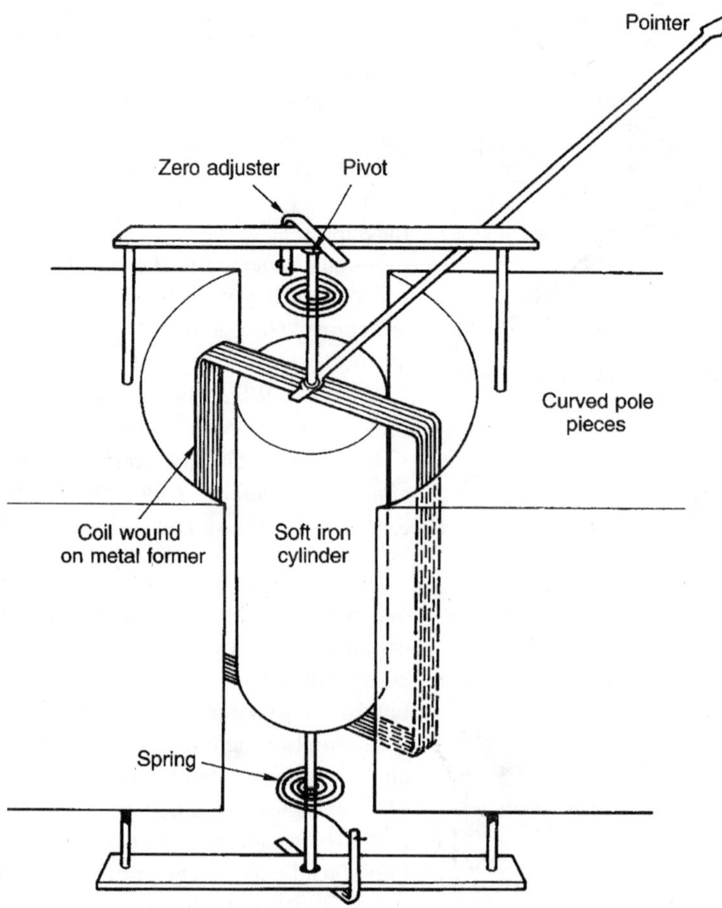

Figure 4.17 Simple sketch of a moving coil galvanometer

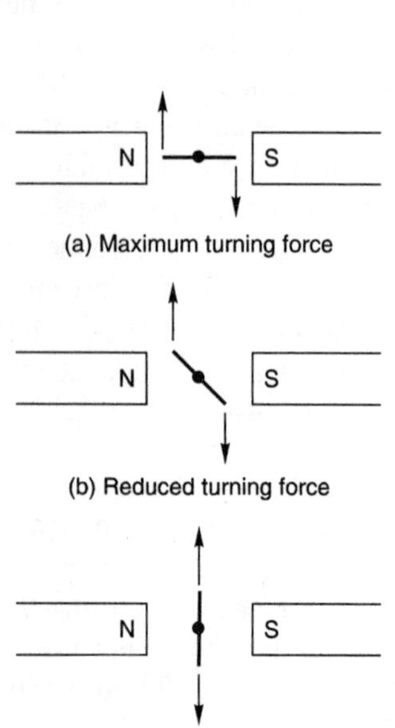

Figure 4.19 The direction of the forces on a current-carrying coil in a magnetic field

In a moving coil meter, the coil is usually wound on a soft iron 'former' and suspended between the poles of the magnet. The current is parallel to the magnetic field at the top and bottom of the coil, so there will be no force there (until the coil turns). The vertical sides of the coil, carrying a current, I, will always be perpendicular to the field so there will always be a force acting.

Remembering the equation $F = BIl$, increasing l will increase the force. This is done quite simply by winding many more turns on the coil; the force is then given by $F = BIl\ (N)$. After this is done, the number of turns, N, remains constant as does the magnetic field provided by the permanent magnet. The only variable, then, is the value of current. The greater the current, the greater the force acting, causing the coil to move. If a pointer is attached to the top of the coil, then it can be made to move over a calibrated scale and measure the current. A sketch of this system is given in Figure 4.17.

Since the coil is in the shape of a rectangle, the current flows round the loop and will travel in opposite directions on either side. This will cause the force to be in opposite directions producing a couple which in turn produces a moment (a turning force), in this special case, called a torque. This is shown in Figure 4.18.

However, the torque depends upon the perpendicular distance between the two forces and this will be reduced as the coil turns in the magnetic field. Hence, as the coil turns, the same force is still applied as the coil remains perpendicular to the field, but the torque reduces. The effect is to produce a non-linear scale on any meter using such a movement. This is shown in Figure 4.19.

The problem is solved by using a radial field. The magnets now have curved pole pieces so that the same force and couple acts, no matter where the coil is in the field. The torque is equal to the value of one of the forces multiplied by the perpendicular distance between them – in this case, the breadth of the coil as shown in Figure 4.20.

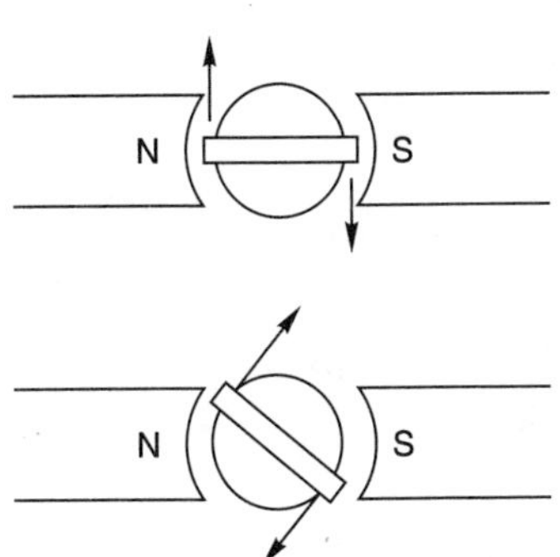

Figure 4.20 The use of curved pole pieces in a moving coil instrument

The motor principle

As with moving coil meters, the basic principle is that a current-carrying conductor in a magnetic field experiences a force.

If a coil is pivoted about its axis such that it is free to rotate and it is carrying a current in a magnetic field, then axial forces will be set up causing the coil to rotate, as we have seen. However, if the coil is to rotate continuously, then some means of allowing the current to flow into and out of the coil must be devised. Furthermore, it must also flow in the appropriate direction to ensure constant torque. The simplest way of allowing current to flow in and out is to use a set of brushes as shown in Figure 4.21.

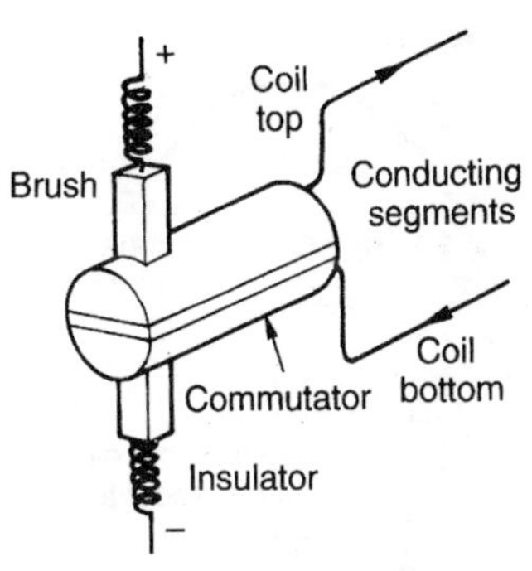

Figure 4.21 Using brushes in a simple electric motor

The brushes come into contact with the commutator which is usually a brass rod split down the middle, the two halves being separated by an insulating material. As the commutator turns it ensures that current in the coil always flows in the correct direction as the commutator changes position with the coil. In small motors the magnetic field, B, can be supplied by a permanent magnet, but in larger machines the main field is supplied by a coil. The motor described is a direct current (DC) motor; in fact, alternating current (AC) motors are much more frequently used since the mains system is AC.

Electric motors are built in a range varying from 1/100 h.p. up to well over 1000 h.p. A 50 h.p. motor is considered a large one. The efficiency of the electric motor is between 75 and 95%. It is higher in large motors than in small ones.

The simple alternator

It has already been shown that when a conductor moves in relation to a magnetic field, an emf is induced (Faraday's first law of electromagnetic induction). A simple form of alternating current dynamo for the continuous production of electric current by electromagnetic induction is shown in Figure 4.22.

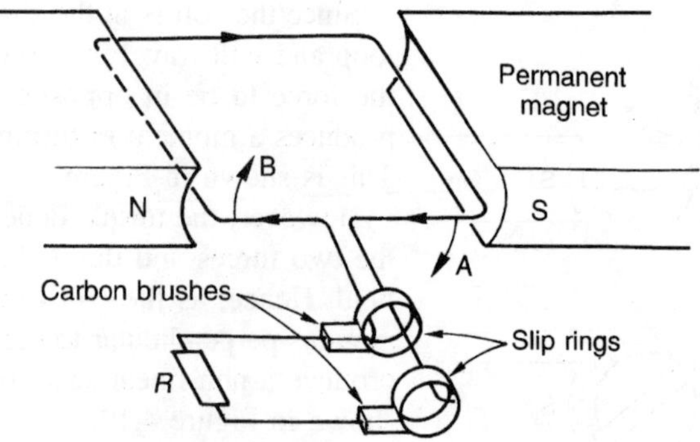

Figure 4.22 A simple alternator for generating an emf

Using slip rings, the emf generated may be connected to a load, shown as a simple resistance, R. Looking now at Figure 4.23, we examine how the alternating current is produced and see that it takes the form of a sine wave.

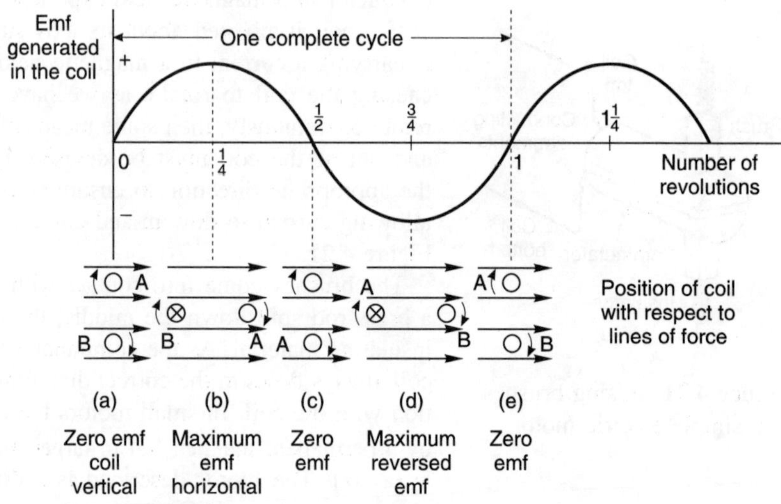

Figure 4.23 The sine wave output from an alternator

Principles of operation of the AC dynamo

1 The rectangular coil is rotated between the poles of a magnet.
2 As the coil rotates its sides cut lines of force and therefore an emf is induced in it. With the coil rotating in a clockwise direction and then using Fleming's Right Hand Rule the direction of the current is as shown.
3 In (a) the sides of the coil are moving along lines of force, there is no cutting of those lines of force so the emf generated is zero.

4 During the first quarter of a revolution the emf increases from zero to a maximum when the coil is in the horizontal position.
5 After this the emf decreases during the second quarter of the revolution and becomes zero when the coil is in the vertical position with side B uppermost.
6 During the second half of the revolution the emf generated follows the same pattern as that in the first half except that the direction of the emf is reversed.
7 If this alternating emf is applied across an external resistance, R, an alternating current (AC) will flow through it.

How to increase the emf from a simple dynamo

Anything which increases the rate of cutting of lines of force will increase the emf.

1 Increase the number of turns on the coil.
2 Wind the coil on a soft iron armature so as to increase the magnetic flux.
3 Increase the speed of rotation.
4 Make the field magnet as strong as possible.

The equation $E = Blv$

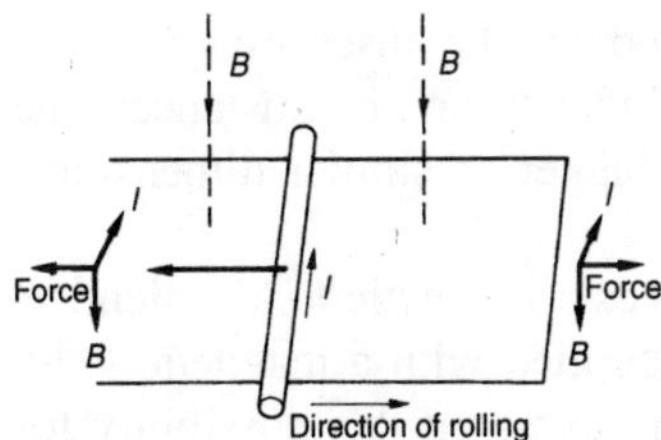

Figure 4.24 How the equation $E = Blv$ is derived

It can be shown that the induced emf is equal to the product of the magnetic field (in tesla), the length of the coil (in metres) and the velocity v (metres per second) at which the coil cuts the magnetic field. To illustrate this, imagine a flat, rectangular piece of rigid copper wire, open at one end, with a metal axle able to roll along it, as shown in Figure 4.24.

The magnetic field is acting vertically downwards; if the metal axle is rolled to the right, an emf will be induced in it in the direction shown which is given by Fleming's Right Hand Rule. If a current flows, then we have a current-carrying conductor in a magnetic field, tending to force the axle to roll in the opposite direction, given by Fleming's Left Hand Rule. As we have already seen, that force, $F = BIl$.

Power is work done per second; since work done (see Chapter 5) is force times distance, we can say that power is equal to force × distance moved per second, and distance moved per second is velocity. Combining this with $F = BIl$ we have:

$$\text{Power} = \text{WD/sec}$$

$$= \text{Force} \times \text{distance moved/sec}$$

$$= \text{Force} \times \text{velocity}$$

Hence: $\text{Power} = BIlv \quad (\text{since } F = BIl)$

$$E \times I = BIlv \quad (\text{since } P = E \times I)$$

Current on each side of the equation is the same and so cancels out, leaving:

$$\underline{E = Blv}$$

At this level you would not be expected to derive the equation (although it is quite logical and not particularly difficult), all that's required is to remember $E = Blv$.

In Figure 4.23 the maximum output voltage which occurs after a quarter of a cycle, three-quarters of a cycle and one and a quarter cycles (and so on) is called the peak voltage. The mains 240 V supply has similar characteristics, the peak value for which is almost 340 V. When we talk about the 240 V mains, we are using what is called the rms value. It's a kind of average value, but if we took the average over one cycle, we would end up with zero which isn't much help. So the values are squared first, which makes the negative values positive, then the mean (or average) value is calculated and square roots taken of the result to obtain the rms value. rms stands for root mean square and the upshot is this equation:

$$\text{rms} = \frac{\text{Peak}}{\sqrt{2}}$$

Electromagnetic induction

We have discussed at length the fact that a current-carrying conductor produces a magnetic field; the magnetism is concentrated if the wire is made into a loop or coil which is often referred to as a solenoid. The magnetic field can be greatly increased by the insertion of a soft iron rod down the centre of the solenoid. Under these circumstances, the coil produces a magnetic field like a bar magnet of similar dimensions, complete with north and south poles.

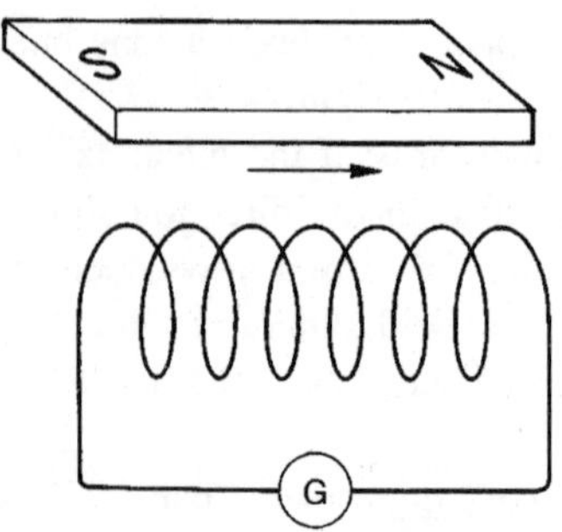

Figure 4.25 Electromagnetic induction

Since a current-carrying conductor produces a magnetic field, scientists thought it logical to ask if a conductor is supplied with a magnetic field, would it produce a current? Michael Faraday explored this possibility for many years. In his notebook dated 17 October 1831, he described how he made a cylindrical coil of copper wire round a paper tube. The ends of the coil were attached to a sensitive galvanometer (which measured the current). Faraday then plunged a magnet into the coil and noticed that the galvanometer needle gave a momentary deflection, showing that current was flowing in the coil. This is shown in Figure 4.25.

The same effect is produced if the coil is moved while the magnet remains stationary. The important point is that there must be relative motion between coil and field so that lines of magnetic force are cut by the coil. Note that it is correct to say that an emf is produced (rather than a current), although when an emf is produced by electromagnetic induction, a current will flow if there is a complete circuit.

In Faraday's early experiments, he plunged a magnet into a static coil. Further experiments led him to realise that it was the changing magnetic field that produced the emf. Since a magnetic field can be produced by a

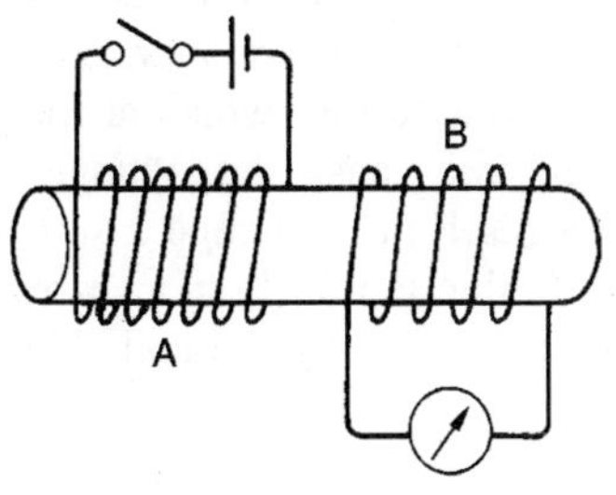

Figure 4.26 Electromagnetic induction without motion

current-carrying conductor, the next step was to plunge a solenoid inside the coil. Sure enough, when current was supplied to the solenoid, the coil surrounding it produced a momentary emf. Having seen the effect, Faraday then turned off the current to the solenoid and saw that the galvanometer registered an emf in the opposite direction. As the current was turned on, the galvanometer kicked one way and when the current was turned off, it kicked in the opposite way. There was no relative motion between solenoid and coil, only a changing magnetic field. The same effect can be produced using the apparatus shown in Figure 4.26.

Here, the coil producing the field (coil A) is placed around a soft iron bar, the other half of which has another coil (coil B) wrapped around it. This is connected to a galvanometer which shows the existence and direction of an electric current. No emf is available from coil B while the battery connected to coil A is disconnected. When it is switched on, a kick is registered on the galvanometer. When a steady current is flowing, no emf is induced. When the switch is opened, the magnetic field collapses and another kick registers in the opposite direction.

What all this means, of course, is that an emf can only be induced in a coil if the magnetic field is constantly varying. Before proceeding further, let's make sure we know what is happening. There are two important points:

1 If a current flows in a coil, a magnetic field is produced.
2 If a coil is placed in a changing magnetic field, an emf is produced and a current flows if there is a complete circuit.

All very straightforward? Good, then here's the next step. In point 2, above, we are saying that an emf can be produced if a coil is in the vicinity of a changing magnetic field. If an emf is induced and there is a complete circuit, a current flows. Now we have the same situation as in point 1; if a current flows in a coil it produces a magnetic field.

Remember that the current flowing in the coil results from the induced emf produced by a changing magnetic field but it still has the same effect. It is not possible to separate the two phenomena; coils carrying a current create magnetic fields, and coils in the vicinity of changing magnetic fields produce currents.

Supposing then, as in Figure 4.25, the north pole of a magnet is plunged into a coil. An emf is induced as indicated by the galvanometer, current flows and a magnetic field is produced. If we arrange for the windings on the coil to be such that the induced emf flows in a direction so as to make it a south pole, if would then attract the magnet to it. Once set up, we could push the magnet towards the coil, induce an emf in it, produce an electromagnet with its south pole at the same end and then let it go. If this really is possible, we should obtain some mechanical energy (provided by the attractive force of opposite magnetic poles) and induce an emf at the same time. Once set up, such a machine could provide unlimited electrical current! Too good to be true? Yes, of course it is.

The problem is that no matter which way round you wind the coil, or which end you present to the moving magnet, as the magnet approaches the coil the induced emf in the coil gives rise to a current which always flows in such a direction as to oppose the motion causing it. Plunge a north pole into the coil and it will produce a north pole; plunge a south pole in and a south pole appears to repel it. You've got to do more work to move the magnet against the repulsive forces set up by the coil. This effect is described by Lenz's law.

Factors affecting the magnitude of the emf

Induced emf increases with:

1 the speed at which the magnetic field changes
2 the area of the coil
3 the strength of the magnetic field
4 the number of turns, N, of the coil.

These results can be formalised in terms of a set of laws. The first two are known as Faraday's laws, the third one is Lenz's law:

1 When a conductor moves in relation to a magnetic field, an emf is induced.
2 The magnitude of the induced emf is proportional to the relative velocity between conductor and field.
3 When an emf is induced in a coil, the current always flows in such a direction as to oppose the 'motion' causing it.

'Motion' is put in inverted commas because, although it is true, the law applies not only if the coil or field are physically moved, but also if the changing field is produced by switching the current to the coil on and off.

Self-induction

Electromagnetic induction takes place because a changing magnetic field induces an emf in an adjacent coil – as we have already seen. However, it is not necessary to have two coils to produce the effect. If a current flows through a coil, it sets up a magnetic field which quickly reaches its maximum value and then remains constant. If the supply of current is interrupted, the field very quickly collapses to zero; during this time the field is changing again so an emf must be induced. Furthermore, it is induced in the same coil that carried the current which set the changing magnetic field up in the first place! The induced emf, by Lenz's law, 'opposes' the motion which caused it. There is no motion as such in this case as we have noted previously; instead, we may regard the induced current as an opposition to the main current as shown in Figure 4.27.

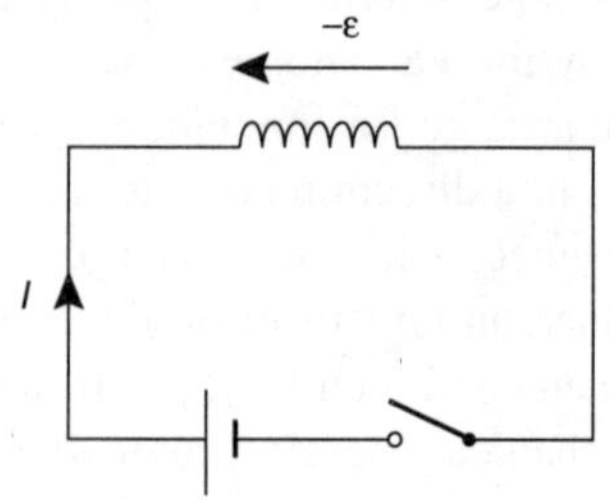

Figure 4.27 Demonstrating self-induction

The emf induced is called a 'back emf' as it opposes the main current in the circuit. The term 'back emf' was mentioned earlier while discussing relay applications. In the touch switch, a diode was included across the relay coil in inverse parallel. We now see that the relay coil produces a changing magnetic field when it is switched off. The change is from B_{max} – the field when the coil carries a current – to B_{zero}, the zero magnetic field when the current is turned off. The change from B_{max} to B_{zero} is extremely rapid; Faraday's second law tells us that the emf is proportional to the rate of change of B and so the emf will be correspondingly large – so large in fact that it could damage the transistor. The inclusion of the diode protects the transistor under these conditions. It should be noted that the back emf is only present while the magnetic field is changing and in a DC circuit this only occurs during switch-on or switch-off.

A back emf is responsible for the spark sometimes seen at switch contacts when a fluorescent light is switched off. This is because such light fittings have a coil inside them to limit the current.

Unit of inductance

The unit of inductance is called the henry(H) in commemoration of a famous American physicist, Joseph Henry (1797–1878), who, quite independently, discovered electromagnetic induction within a year after it had been discovered in England by Michael Faraday in 1831. In radio work the units mH and μH are common. The symbol for inductance is L (in the same way that R stands for resistance and C for capacitance). The circuit symbols are as shown in Figure 4.28.

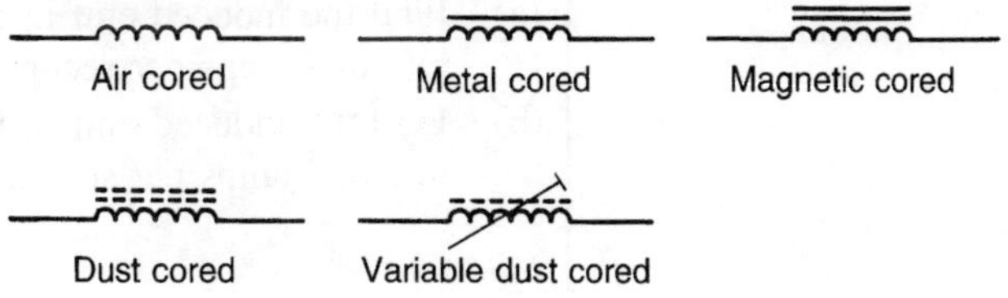

Figure 4.28 Circuit symbols of inductances

Definition

A coil has an inductance of 1 henry if an emf of 1 volt is induced in the circuit when the current changes at the rate of 1 ampere per second. This can be represented by a simple mathematical equation; however in so doing, we introduce the Greek letter D or delta, which is shown as a small triangle Δ and which means 'a change of':

$$L = \frac{E}{\Delta I/t}$$

The minus sign indicates a back emf. So, in the equation, L is the inductance, E is the back emf, ΔI is the change in current and t is the time during which the change takes place, hence '$\Delta I/t$' means 'the rate of change of current'.

It is unlikely that you will be expected to use the equation at this level, you simply need to know the relationship.

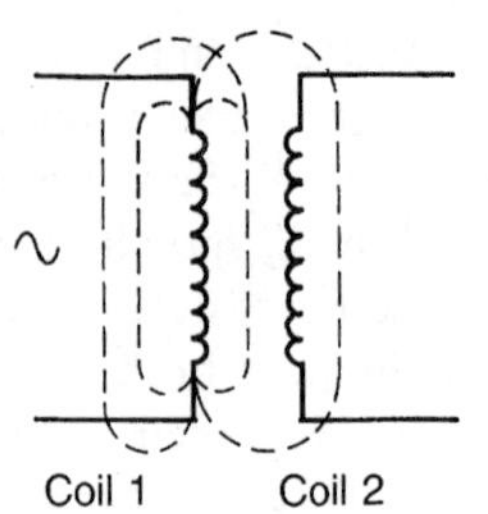

Figure 4.29 Demonstrating mutual inductance

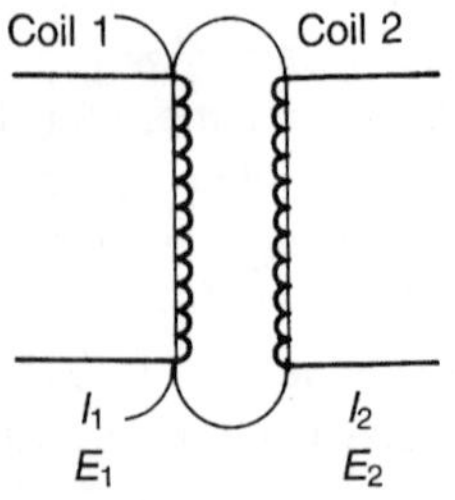

Figure 4.30 Two coils experiencing mutual inductance

Mutual inductance

If two coils are placed close to each other as shown in Figure 4.29, such that the magnetic field of one coil links the turns of the other, then an emf will be induced in the second coil. The magnetic field is created by a changing current in one coil, inducing an emf in the other.

Turning to Figure 4.30, the induced emf in coil 2 depends upon the changing magnetic field (or flux) in coil 1. The changing flux depends upon the current flowing per second so it follows that the induced emf in coil 2 depends upon the rate of change of current in coil 1. The symbol for mutual inductance is M, the unit, as for self-inductance, is the henry, H.

If the current in coils 1 and 2 is I_1 and I_2 respectively, and the emfs are E_1 and E_2 respectively, then:

$$E_1 = M\frac{\Delta I_2}{t} \quad \text{and} \quad E_2 = M\frac{\Delta I_1}{t}$$

Example 4.1

The mutual inductance between two coils P and S is 5 H.

(a) Find the induced emf in S while the current in P changes at the rate of 2 amps per second.

(b) Find the induced emf in P while the current in S changes at the rate of 5 amps per second.

Solutions

(a) Using the equation: $E_1 = M\frac{\Delta I_2}{t}$

hence: E_1 or $E_S = 5 \times 2 = 10\,\text{V}$

(b) Similarly: $E_p = 5 \times 5 = 25\,\text{V}$

Transformers

The effect of mutual inductance is the basis of operation of the transformer. The coil to which the changing current is applied is called the primary and the coil which has the emf induced in it is called the secondary. The mutual inductance of two linked coils depends upon the number of turns in each; in a transformer, the ratio of the number of turns in the secondary to the number of turns in the primary is called the turns ratio. Hence, using the symbols N_1 and N_2 for the number of turns in primary and secondary respectively, we can write:

Turns ratio $= N_2/N_1$

The mutual inductance of two linked coils also depends on the sizes of their coils, their proximity to each other and the medium linking them. In the absence of anything else, air would be the medium but the flux linkages would be rather poor. Iron is often used because it increases the flux linkages many hundreds of times. However, the flux linkages in the iron itself also set up induced emfs which flow as 'eddy' currents. By Lenz's law, these small, circulating currents tend to oppose the 'motion' producing them. For this reason, transformer cores are laminated. This means that instead of the core being a solid iron block, it is made up from many thin, flat pieces, interleaved together. This is shown in Figure 4.31.

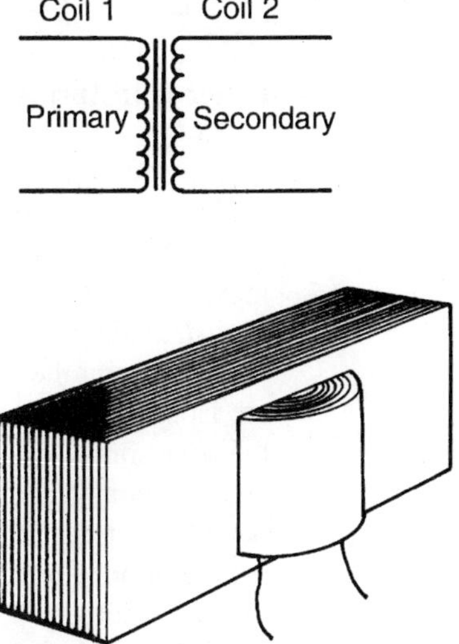

Figure 4.31 Two coils linked by a laminated core

Remembering that an emf is induced only when there is a changing magnetic field, any direct current fed to the primary of a transformer would have to be repeatedly turned on and off. Since an alternating current is continuously passing through zero, it is obvious that any transformer would be working under ideal conditions when fed with alternating current. This is one of the main reasons why the mains 240 V supply in our homes is alternating current (AC).

Transformers tend to be extremely efficient, that is to say, the total power supplied at the primary is available at the secondary. However, there will be some losses because the laminated core heats up as it is being continually magnetised and demagnetised by the varying current supplied to it. Even so, efficiencies can be expected to exceed 95% in most cases and often are greater than that. For simple calculations, we usually assume 100% efficiency, in which case, the ratio between the emf

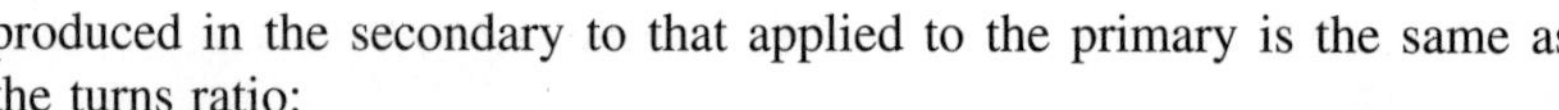

produced in the secondary to that applied to the primary is the same as the turns ratio:

$$\frac{E_2}{E_1} = \frac{N_2}{N_1}$$

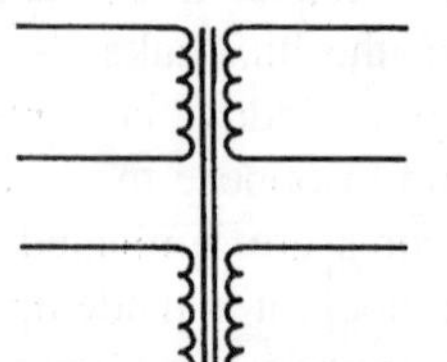

Figure 4.32 A transformer with centre-tapped windings

For a perfect voltage transformer, the ratio of the output emf to the input emf equals the turns ratio of the coils. Many transformers have split windings in order to make them more versatile. For example, a transformer designed to be operated from the mains supply to produce 12 V on the secondary, having split windings on both primary and secondary, may be used on the UK 240 V mains by connecting the primary windings in series, to produce 12 V on the secondary by connecting the windings in series, or 6 V on the secondary by connecting the windings in parallel. This is shown in Figure 4.32.

Other transformers

It is instructive to look at manufacturers' catalogues in order to obtain further examples of the many transformer configurations which are available. Those described so far, with separate primary and secondary windings, are called 'double wound' transformers. This means that there is no *electrical* connection between the two coils, the emf induced in the one resulting from the changing magnetic field in the other. Since the two coils are isolated, all these transformers are known as isolating transformers. In the past, transformers with the same number of windings on both primary and secondary have been produced; the voltage on each side is the same, but the fact that they are electrically isolated means that anyone using such a supply has protection from the mains as the current available is thereby limited.

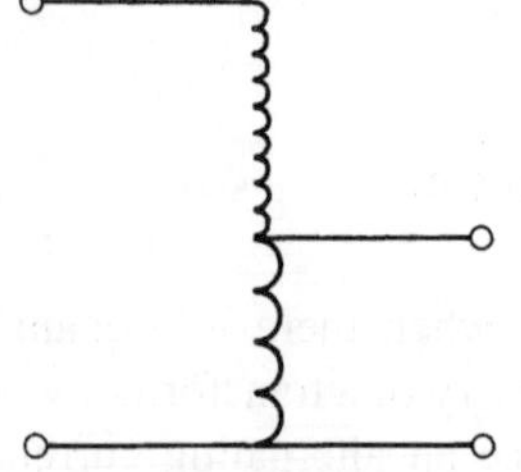

Figure 4.33 Circuit symbol of an auto transformer

Auto transformers

There are some transformers, however, which have only one winding and therefore do not provide any isolation from the mains; they are called auto transformers. The advantage of these types is that they are physically smaller, lighter and cheaper than their double wound counterparts. The circuit symbol for an auto transformer is shown in Figure 4.33.

Multiple choice questions

1 The correct units for force, magnetic field, current and length, respectively, are

A kilograms, tesla, amps, centimetres
B newtons, tesla, amperes, centimetres
C newtons, webers, amperes, metres
D newtons, tesla, amperes, metres

2 The correct ‘force on a current-carrying conductor’ equation is

A $F = BIl$
B $B = FIl$
C $I = FBl$
D $l = FBI$

3 A straight wire of length 3 m is placed perpendicular to a magnetic field, B. If a force of 6 N acts on the wire when a current of 200 mA flows through it, the value of B is

A 0.01 T
B 0.1 T
C 3.6 T
D 10 T

4 A straight wire of length 20 cm is situated in a uniform field of 0.5 T carrying a current of 2 A. The force acting on the wire is

A 0.2 N
B 2.0 N
C 20 N
D 20.2 N

5 Electromagnets usually have their coils wound on cores of

A copper
B cobalt
C alnico
D soft iron

6 The direction of the magnetic field surrounding a current-carrying conductor can be worked out using

A Fleming’s Left Hand Rule
B Fleming’s Right Hand Rule
C Maxwell’s corkscrew rule
D Lenz’s law

7 The only metal from the list below which is *not* magnetisable is

A cobalt
B copper
C iron
D steel

8 A straight wire of length 10 cm is placed at right angles to a uniform field of intensity 0.4 T. If the force acting on the wire is 0.2 N, the current, I, is

A 0.5 A
B 0.8 A

C 2.5 A
D 5.0 A

9 From the list below the metal which produces the strongest magnetic field when used in an electromagnet is

A aluminium
B copper
C soft iron
D nichrome

10 A small, straight conductor of length 10 cm, suspended at right angles to a uniform field, *B*, experiences a force of 0.0015 N. If the current flowing is 3 A, the field strength will be

A 0.005 T
B 0.05 T
C 0.5 T
D 3.5 T

11 A small aluminium rod placed parallel to a magnetic field of 1.5 T has a length of 2 m and carries a current of 3 A. The force acting on it is

A zero
B 0.45 N
C 0.90 N
D 9.0 N

12 A 'step up' transformer is one which has

A more turns on the primary than on the secondary
B more turns on the secondary than on the primary
C a larger primary voltage than secondary voltage
D a larger secondary current than primary current

13 The magnetic field of a transformer will be increased the most by having a core of

A soft iron
B cobalt
C copper
D aluminium

14 The symbol shown in Figure 4.34 represents

Figure 4.34

A an auto transformer
B a mains transformer
C an inductor with an iron core
D an inductor with a ferrite core

15 Mutual inductance is given the symbol

A I
B M
C L
D B

16 Two high voltage coils have a mutual inductance of 15 H. If the current changes at the rate of 100 A/s, the back emf is

A 0.15 V
B 6.67 V
C 150 V
D 1500 V

17 The symbol shown in Figure 4.35 represents

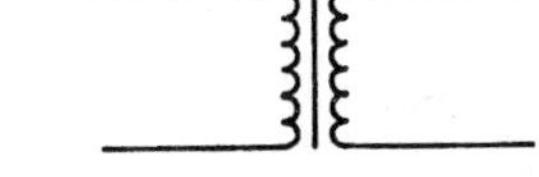

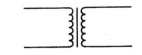

Figure 4.35

A a single wound transformer
B a double wound transformer
C an auto transformer
D a variac transformer

18 A transformer has 1600 turns in the secondary; when a voltage of 16 V is applied to the primary, an emf of 80 V is induced in the secondary. The number of turns in the primary is

A 320
B 3200
C 800
D 8000

19 The symbol shown in Figure 4.36 is

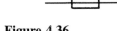

Figure 4.36

A a resistor
B a coil
C a fuse
D a lamp

20 A mains choke of inductance 10 H produces a back emf of 900 V when the current reduces from 3 A to zero. The time taken for the current to change is

A 0.033 s
B 0.33 s
C 0.27 s
D 270 s

21 The symbol shown in Figure 4.37 represents

Figure 4.37

A an iron cored inductor
B an air cored inductor
C a dust cored inductor
D an inductor with a variable dust core

22 The basis of operation of the transformer is the effect of

A self-inductance
B mutual inductance
C coupled inductance
D decoupled inductance

23 The 5000 turn primary of a transformer is fed with 100 V. Assuming 100% efficiency, if the secondary has 50 turns the output voltage will be

A 0.1 V
B 1.0 V
C 100 V
D 10 000 V

24 An auto transformer has

A more turns on the primary than on the secondary
B more turns on the secondary than on the primary
C more than one winding
D only one winding

25 Auto transformers are used in some applications because

A they are safer
B they are smaller
C they are cheaper
D they produce no interference

26 'When an emf is induced in a coil, the current always flows in such a direction as to oppose the "motion" causing it.' This law is known as

A Faraday's law
B Henry's law
C the induction law
D Lenz's law

27 The high back emf produced when current is interrupted in a coil is used in

A a filament lamp
B a light-emitting diode
C a fluorescent light
D a neon lamp

28 Magnets with curved pole pieces are used in a moving coil galvanometer because

A they produce a stronger magnetic field
B they allow the instrument to have a linear scale
C they are easier to manufacture
D it is less likely that the coil will touch the magnet

29 In a loudspeaker, the current-carrying conductor is called

A an electric coil
B a magnetic coil
C a speech coil
D the cone

30 The moving part in a relay is called

A an armature
B a slip ring
C the yoke
D a moving coil

5 Heat and mechanical units

Heat transmission

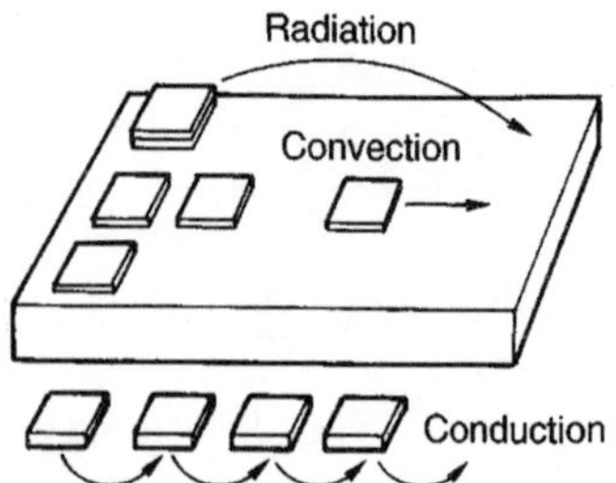

Figure 5.1 Conduction, convection and radiation

There are three methods of conveying heat and they can be described effectively using the following analogy: imagine a long table with a pile of books at one end, as shown in Figure 5.1, then consider how you can get all the books from one end of the table to the other.

There are three obvious possibilities:

(a) Involve a group of people spanning the length of the table and have them pass the books along like a bucket brigade.
(b) Pick up each book, one at a time, and carry them to the other end of the table.
(c) Simply pick them up one at a time and throw them to the other end of the table.

These three ways of transporting books are meant to correspond to the three methods of heat transfer: (a) conduction, (b) convection, and (c) radiation. Conduction occurs in solids and liquids where the application of heat causes the molecules to vibrate more rapidly and this increased energy is passed on down the length of the solid or liquid. Conductors of heat tend also to be good conductors of electricity since in each case a similar mechanism is involved. You may recall from Chapter 1 that the heat of a body is related to the motion of the particles from which it is made.

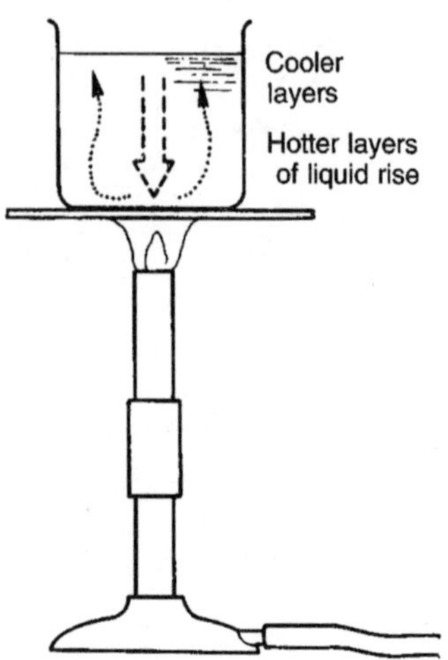

Figure 5.2 Convection currents in heated water

Convection occurs in liquids and gases. When substances are heated they tend to increase in volume which therefore reduces their density; this in turn causes their bulk to 'rise' since less dense substances float on top of more dense substances (imagine a cork floating on water, for example). Convection currents are set up in a liquid if it is heated from the bottom. The liquid in closest contact expands, its density reduces and so it 'floats' to the top. This is shown in Figure 5.2 with the classic laboratory icon of a beaker of liquid on a tripod being heated by a Bunsen burner.

As the warm water rises, cooler (more dense) water falls to take its place. When the warmer water reaches the top of the beaker, it cools down again and starts to sink and is itself replaced by more warmer water from the bottom. This process continues as long as the liquid is being heated and there is a difference in temperature between the upper and lower layers of the liquid – convection currents are set up.

All fluids (the word covers liquids and gases) can have convection currents set up in them and this is put to good use in the cooling of electronic circuits. Provided the devices generating the most heat are placed suitably, they will heat the surrounding air and if this air is allowed to rise and escape, cooler air will flow in at the bottom to replace it. As the cycle continues, the device is successfully kept cool.

Effect of heat on resistance

Power output transistors may develop a considerable amount of heat which has to be disposed of. As a matter of common sense most things expand when heated. This can now be taken a stage further by stating that the electrical resistance of metals also tends to increase with an increase of temperature. The amount by which the resistance of a conductor increases with an increase in temperature is known as the temperature coefficient of resistance.

If the resistance of the metal increases with an increase in temperature then it is said to have a 'positive temperature coefficient' (PTC). Most metals are PTC devices and there are many electronic components which exploit this phenomenon for temperature sensing and measurement – more about this later.

Meanwhile, transistors are generally made from silicon, and this has a negative temperature coefficient of resistance (NTC). In other words, the hotter it gets, the more the resistance reduces. If the resistance reduces, more current can flow and if more current flows, the device becomes even hotter and so the resistance reduces even more and so even more current flows – it's a vicious circle. In electronics this is called thermal runaway, and this cycle of events is illustrated in Figure 5.3.

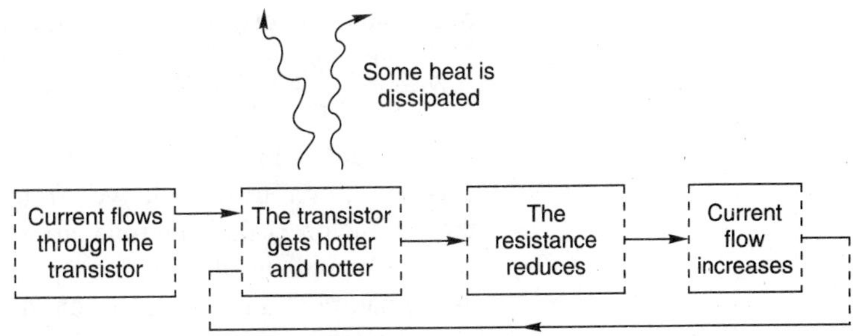

Figure 5.3 The onset of thermal runaway

By placing the transistor (or other device) on a heat sink – simply a piece of metal with a large surface area – the heat is quickly conducted away, the surrounding air then sets up convection currents and the device is kept cool. Heat sinks often consist of a series of 'fins' in order to increase the surface area without increasing the overall size too much.

Radiation

The third method of heat transfer is through radiation. Radiated heat (infrared radiation) is a form of electromagnetic energy (the same as radio waves and light itself) the frequency of infrared rays being slightly lower than visible red light. Experimentation has shown that radiation from metal surfaces takes place more efficiently if the surface is dull; a matt black finish on a heat sink therefore increases the amount of heat being dissipated. This is the reason why motor car radiators, which are generally made from brass, are painted black, and electric kettles are usually made from silvery, or chrome, polished metal with a shiny surface. This reduces the amount of heat radiated and leaves more to heat the water in the kettle – the efficiency of the device is increased.

In summary, you should note that heat sinks are made from metal because it is a good heat conductor, the surface area is made as large as possible using fins to produce convection currents in the surrounding air, and they are often painted a dull black to increase heat dissipation through infrared radiation. In many home computers a fan is also installed in order to provide a larger flow of air to dissipate the heat generated in the power supply.

Temperature coefficient of resistance

It is only fair to point out that much of this section is included for the very good reason that it is in the syllabus and it's also good reference material. However, the level of understanding required is minimal and, as always, is reflected in the type of questions you may come across in the exam, and some specimen multiple choice questions are, as always, given at the end of the chapter.

The temperature coefficient of resistance of a material is the increase in resistance of a specimen expressed as a fraction of the resistance at 0° per K rise in temperature. (K stands for Kelvin, the unit of temperature on the scientific, thermodynamic scale. In practice, it means much the same thing as the familiar degrees Celsius.) The temperature coefficient of resistance is given the symbol α (alpha), hence:

$$\alpha = \frac{R - R_o}{R_o t} \qquad \text{or } R = R_o(1 + \alpha t)$$

When a metal is heated, its electrical resistance increases. Remembering Chapter 1, current flow derives from an orderly drift of free electrons in the conductor. If the electrons are heated, they move more randomly and with increased speed, resisting the tendency to drift in an orderly fashion as the nature of an electric current dictates. This is why what is called the electrical resistance of a metal increases with an increase in temperature.

Example 5.1

The resistance of an iron wire which is 5 Ω at 0°C becomes 6 Ω at 50°C. Calculate the temperature coefficient of resistance of the iron. Note that 't' in the equation is the temperature *change*, in this case $50 - 0 = 50$.

Solution

From $R = R_0(1 + \alpha t)$ $\quad 6 = 5(1 + 50\alpha)$

$$6 = 5 + 250\alpha$$

$$250\alpha = 6 - 5$$

so: $\alpha = 1/250 = 4 \times 10^{-3}\ \text{K}^{-1}$

Example 5.2

An aluminium bar has a resistance of 1 Ω at 0°C. Calculate its resistance at 88°C given that the temperature coefficient of resistance of aluminium is $4 \times 10^{-3}\ K^{-1}$

Answer
1.352 Ω

Wirewound resistances were mentioned earlier; they are used when the current flowing through them is quite high. Naturally, the components become very hot in normal use, so it is important to use a wire that does not change in resistance significantly with an increase in temperature. Example 5.3 demonstrates this:

Example 5.3

A standard resistance made from manganin wire has a resistance of 10 Ω at room temperature (say 20°C). Calculate its resistance at 100°C.

Solution

$$R_t = R_o(1 + \alpha t)$$
$$= 10(1 + 10^{-5} \times (100 - 20))$$

so: $R_t = 10.008\ \Omega$

In other words, it has hardly changed at all, which is precisely what is required. What would be the resistance at the higher temperature if the resistor were made from mild steel wire ($\alpha = 50 \times 10^{-4}$/K)?

Solution

$$R_t = 10(1 + (5 \times 10^{-3} \times 80))$$

so: $R_t = 14\ \Omega$ – a change of 40%

Example 5.4

In the previous example, what would the resistance value be at the higher temperature if the resistor were made from copper wire, $\alpha = 3.9 \times 10^{-3}\ K^{-1}$?

Solution

$$R_t = 10(1 + (3.9 \times 10^{-3} \times 80))$$

so: $R_t = 13.12\ \Omega$ – a change of over 30%

Example 5.5

If a piece of copper of resistance 1 Ω at 0°C has its temperature progressively reduced, the resistance will decrease until it eventually drops to zero. At what temperature will this occur? (For copper, $\alpha = 3.9 \times 10^{-3}\,\mathrm{K}^{-1}$.)

Solution

Using $R_t = R_o(1 + \alpha t)$ and rearranging:

$$t = \frac{R_t - R_o}{\alpha R_o} = \frac{0 - 1}{\alpha \times 1}$$

so: $$t = \frac{-1}{3.9 \times 10^{-3}} = -256.4°\mathrm{C}$$

The value −256.4°C is about 16°C above absolute temperature. In fact, the change in resistance per °C is not constant below about 50°C so the figure is approximate. However, the calculation indicates that very low resistance may be attained at extremely low temperatures and this is the principle behind 'superconductors'.

Thermistors

Devices which have very large changes in resistance with a change in temperature are used in electronics. They are called thermistors, derived from the term 'thermal resistor'. These components respond to temperatures over useful ranges and there are many different types and values. The resistance of wirewound resistors increases as the temperature increases but usually not very much as we have seen. But thermistors are designed to have significant changes in resistance with a change in temperature.

There are two types of thermistor, those which increase in resistance with an increase in temperature, and those which decrease in resistance with an increase in temperature. The former are called positive temperature coefficient (PTC) devices and the latter negative temperature coefficient (NTC). NTC types are the most common where the resistance diminishes rapidly with an increase in temperature. There are various symbols for thermistors and these are shown in Figure 5.4.

Physical construction

Basically, every thermistor application is one of temperature measurement or the detection of temperature changes. The devices can be heated externally or by the action of a current passing through them. The rod type of thermistor may have a resistance of about 200 Ω at room temperature (about 20°C) which will reduce to 5 or 6 Ω at about 100°C. These, as well as disc and bead type thermistors, are shown in Figure 5.5.

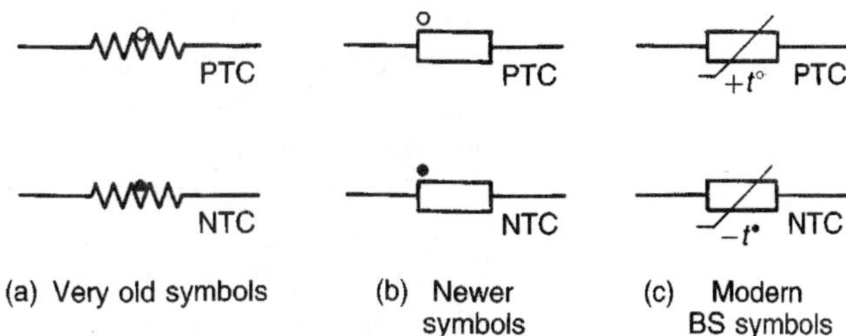

Figure 5.4 Thermistor symbols

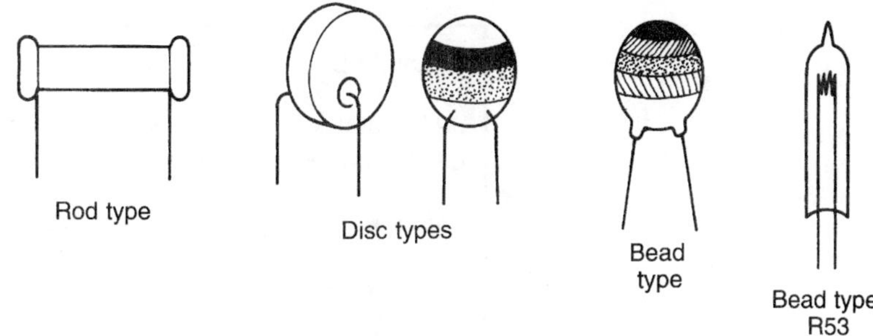

Figure 5.5 Common types of thermistor

Mechanical units

The story goes that an Englishmen talking to an Irishman said, 'I've heard that Irishmen always answer a question with a question,' and the Irishman said, 'Who told you that?' I slip that little anecdote in for two reasons: one, as soon as the words 'mechanical units' are mentioned, your average student switches off totally and I'd like to avoid that; and two, I now present a question that really does require an answer.

What's the difference between mass and weight?

A good response might be that mass is the quantity of 'stuff' in a body; weight is related to the force of gravity acting on a mass. The Earth's gravity varies slightly over its surface but a good approximation is 9.81 N/kg. (This may be remembered by noting that 9^2 is 81). 'N/kg' is pronounced 'newtons per kilogram'. For most purposes, 10 N/kg is close enough.

So on Earth, a 1 kg mass has a weight of 9.81 N; in space, where there is no gravity to speak of, a mass may become weightless – it happened to Neil Armstrong and Co. when they were on their way to the Moon in 1969. Speaking of which, a 1 kg mass is still a 1 kg mass on the Moon, but it weighs less because the Moon's gravity is less, being about one-sixth that of the Earth, amounting to approximately 1.6 N/kg.

In the UK we tend still to talk about our 'weight' in stones and pounds, but this is incorrect. Stones and pounds are units of mass; weight is a force and so we should really quote our weight in newtons. A good guide is to note that what we used to call a 2 lb bag of sugar is now a 1 kg bag of sugar, 1 kg being equal to about 2.2 lb. A 1 kg bag of sugar therefore weighs 9.81 N.

In summary, then, mass is measured in kilograms (kg), force (and weight, since it is the force on a body due to gravity) is measured in newtons (N). Work done is equal to force times distance moved in the direction of the force, or WD = force × distance. Energy is the ability to do work and both work done and energy are measured in joules. Power is the rate of doing work, which is work done per second (joules per second, or watts) or force times velocity.

Example 5.6

A 1 kg mass has a force of approximately 10 N acting on it, due to the force of gravity. What is the weight of a 50 kg mass?

Solution

$$\text{Weight} = \text{mass} \times \text{force of gravity acting on it}$$

$$\text{so: Weight} = \text{mg} = 50 \times 10 = 500\,\text{N}$$

Example 5.7

What is the work done in moving a 50 kg mass through 15 m?

Solution

$$\text{WD} = \text{force} \times \text{distance} = 500 \times 15 = 7500\,\text{joules}$$

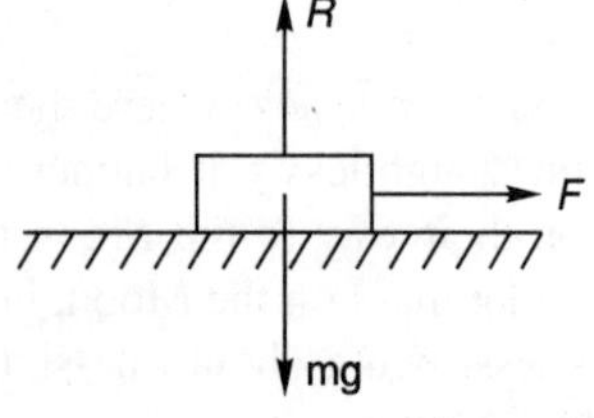

Figure 5.6 Work done in moving an object

Note If a mass is pulled along a surface, some force will be required to overcome friction (where it exists), so it is usual to assume that the force used to pull the object is the net force. Further, the definition of work done particularly specifies that the direction of movement is the same as the direction of the force. If it isn't, the net force will be reduced.

Figure 5.6 shows a mass sitting on a horizontal plane; its weight, mg, is acting downwards and if the mass is stationary, there must be an equal and opposite force acting upwards which is called the normal reaction, *R*. The force being applied to move the mass is marked *F*.

If the mass is being pulled at an angle, it is possible to identify two components of the force. One which acts in the direction of movement (horizontally) and one which acts vertically. In this case the net force causing movement is reduced.

What is of interest is that if you pull something like a heavy roller used on a cricket pitch, it is easier to pull it than to push it, because part of the force applied is pulling the roller off the ground and the rest is pulling it along. If you push the roller, part of the force is pushing it into the ground.

The same applies to supermarket trollies – in theory anyway. In practice, however, as you may have noticed, supermarket trollies have a mind of their own, whether pushed or pulled, and in the hands of some shoppers are downright dangerous!

Waves and sound

Wave motions can be divided into two groups, mechanical and electromagnetic. Mechanical waves consist of particles which are vibrating, whereas electromagnetic waves consist of varying electric and magnetic fields. The former can vary in speed of propagation, but all electromagnetic waves have the same speed in a particular medium, which is the speed of light, approximately 3×10^8 ms^{-1} in air.

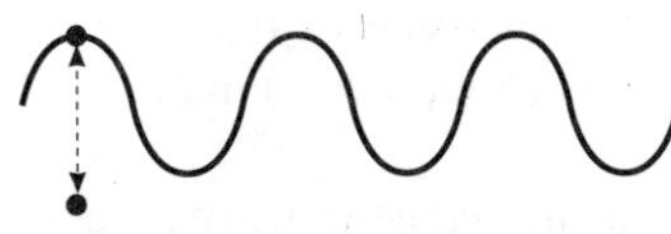

Figure 5.7 Energy, not matter transference, of waves on water

If a stone is thrown into a pool, waves are set up; these are called *transverse* waves because the wave motion is at right angles to the direction of the wave. It is important to note, however, that it is the energy of the wave that moves and not the water itself, although the water does move up and down as shown in Figure 5.7.

Definition

A wave is defined as a disturbance, usually periodic, which travels with finite velocity through a medium and remains unchanged in type as it travels. Mechanical waves may be divided into two categories: longitudinal waves and transverse waves. Imagine a train moving off; the trucks are pulled along from rest so when they first experience a force they accelerate.

Longitudinal waves

Note that an applied force causes acceleration, so that in our example, the trucks' speed increases. You don't need to increase the force to make the trucks go faster. Between each truck are buffers which cause all the trucks to settle down to the same speed, so they oscillate a bit at first until that speed is reached. Sound waves are propagated in the same way. The density of the air is varied by successive compressions and rarefactions as shown in Figure 5.8.

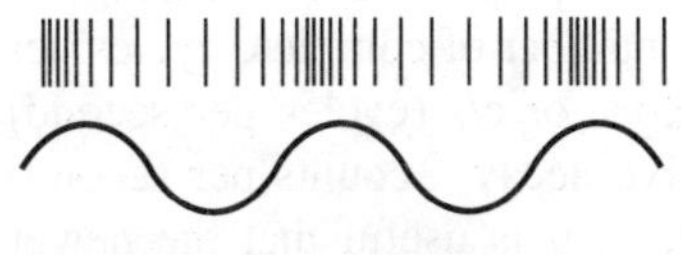

Figure 5.8 Longitudinal waves produced by sound in air

A 'slinky' spring is a good way of demonstrating the effect. The important point to note is that longitudinal wave motion takes place in the same direction as the direction of propagation of the wave. Compare this to a transverse wave whose motion is at right angles to the direction of propagation.

Note that mechanical waves need a medium in which to travel, but electromagnetic waves do not. For mechanical waves, the greater the density of the medium, the more quickly the waves propagate. For electromagnetic waves, the reverse is true.

Table 5.1 summarises the above and gives some examples of each type of wave motion.

Table 5.1 Summary of wave motions with examples

WAVES	
Mechanical	*Electromagnetic*
Need a medium to travel in, e.g. air, water, metal	Can travel in a vacuum
Can be detected (heard) by the human ear	Cannot be heard
Travel at about 330 ms^{-1} depending upon medium and temperature	Travel at a speed of light, c, which is about 3×10^8 m/s
May be either longitudinal or transverse	All electromagnetic waves are transverse

EXAMPLES		
Transverse	*Longitudinal*	*Transverse*
Waves on a pond	Sound waves	Radio waves
Guitar string	'Slinky' spring	X-rays
Record pickup	Train moving off	Light
Sea waves		Microwaves
Wind-swept field of corn		Infrared radiation – radiated heat

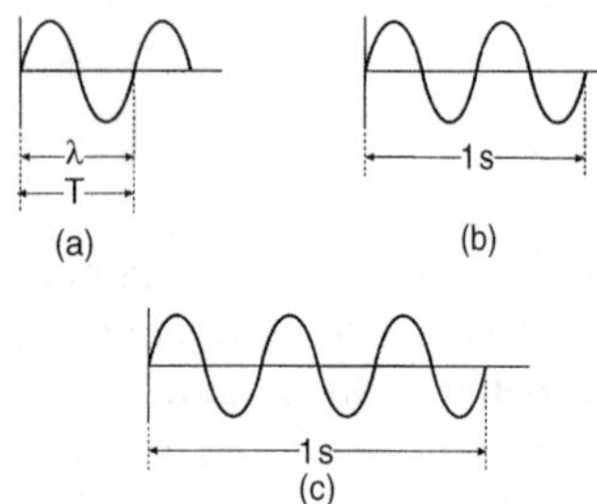

Figure 5.9 (a) One wavelength. (b) A frequency of 2 Hz. (c) A frequency of 3 Hz

Wavelength and frequency

The frequency of a wave motion is the number of complete vibrations per second, for example one complete 'to and fro' or 'up and down' movement. One complete vibration travels through a cycle, ending up where it started. Frequency is therefore the number of complete cycles per second. The old unit for frequency was c.p.s. or c/s (cycles per second) but c.p.s. is also used in physics (radioactive decay – counts per second) and in computing (characters per second), so it is useful that the newer unit of hertz (Hz) is now in common use.

Figure 5.9 helps us to describe wavelength and frequency. In Figure 5.9(a), the wavelength – literally the length of one wave or complete cycle – is marked. Wavelength is given the symbol λ and is measured in metres. Suppose it takes one second to complete two cycles (see Figure 5.9(b)) the frequency is then two cycles per second or 2 Hz.

If it takes one second to complete three cycles, the frequency is obviously 3 Hz. The period of the waveform is the time taken to complete one cycle, so if the frequency is 3 Hz, the period, T, is 1/3 s. So $f = 1/T$ and $T = 1/f$.

Velocity is equal to distance travelled divided by the time taken. For a wave, the distance travelled is the wavelength and the time taken is the period, T.

$$\text{Velocity} = \frac{\text{Distance moved}}{\text{Time taken}} = \frac{\text{Wavelength}}{\text{Period}}$$

but: Period, $T = 1/f$ so: $v = \dfrac{\lambda}{1/f}$

and: $\underline{v = f\lambda}$

For light, in air, $\underline{v = 3 \times 10^8\ \text{ms}^{-1}}$

For sound, in air, $\underline{v = 330\ \text{ms}^{-1}\ \text{(approx.)}}$

Example 5.8

The speed of light is $3 \times 10^8\ \text{ms}^{-1}$. Calculate the wavelength of a radio wave whose frequency is 450 kHz.

Solution

$V = f\lambda$ so: $\lambda = \dfrac{v}{f}$

$$\text{and: } \lambda = \frac{3 \times 10^8}{450\,000} = 667\ \text{m}$$

Light and optics

Light waves travel in straight lines – we cannot see round corners! This is called 'rectilinear propagation of light'. Although Einstein (famous for his $e = mc^2$ equation) stated that light bends under the influence of gravitational fields, we are here only concerned with Earth-bound observations. Unfortunately, the true nature of light is still not properly understood and scientists talk about the 'wave/particle duality of light'. What this means is that certain experiments seem to 'prove' that light consists of particles called photons, and other experiments seem to 'prove' that light is a wave motion. Fascinating though this is, we limit our discussion of light to more modest observations as follows:

1 Light travels in straight lines.
2 Light is a form of energy.
3 Light has different velocities in different media. Its maximum velocity is in a vacuum. The greater the optical density, the less the velocity of light.
4 A ray of light is reversible, that is to say, if a ray of light is shown to take a particular path through a system of lenses or via mirrors, the path taken will be exactly the same, no matter which direction it travels in.

Reflection at plane surfaces

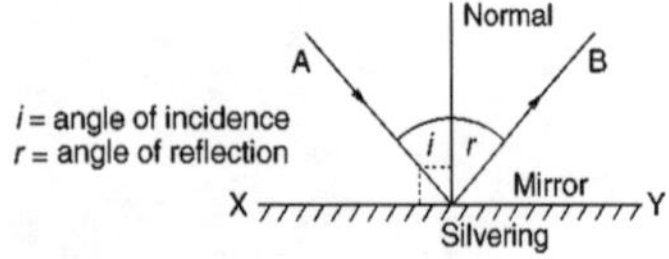

Figure 5.10 Reflection at plane surfaces

Figure 5.10 shows a single ray of light being reflected by a plane mirror. The angle of incidence is denoted 'i' and the angle of reflection is denoted 'r'.

Experiment shows:

1 The reflected ray, the incident ray and the normal to the mirror at the point of incidence, all lie in the same plane.
2 The angle of incidence equals the angle of reflection.

These are known as the two laws of reflection. As a consequence of the first law, a ray incident at a given angle on the surface is reflected in a definite direction which can be predicted. The reversibility of light means that if light travels in the opposite direction it would still travel along the same path.

Refraction

If light shines into a more dense medium – for example, from air into water or from air into glass – the rays of light bend. Travelling from a less dense medium to a more dense one, the ray is bent towards the normal. As the ray of light leaves the more dense medium, it is bent away from the normal. This is shown in Figure 5.11.

There is a constant relationship between angle of incidence and angle of refraction for a given pair of media which is known as Snell's law:

$$\frac{\sin i}{\sin r} = n$$

What this means is that if you take the sine of the angle of incidence and divide it by the sine of the angle of refraction, the value will always be the same for a given medium, for example glass, although there are many different types of glass, all having their own value of n. Note that, unlike angles of incidence and reflection for rays incident on a mirror (which are equal), angles of incidence and refraction are *not* equal.

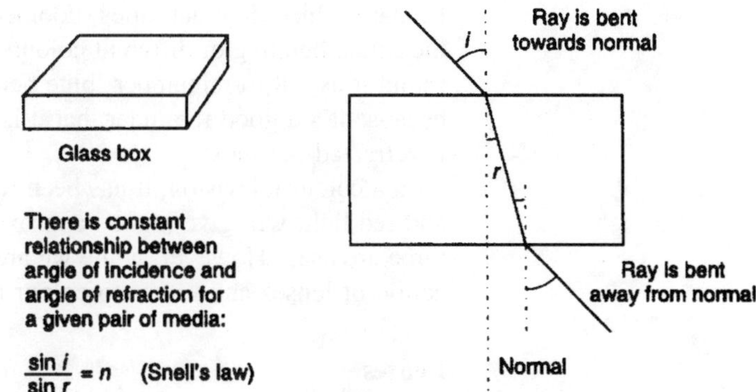

Figure 5.11 Refraction of light in a glass block

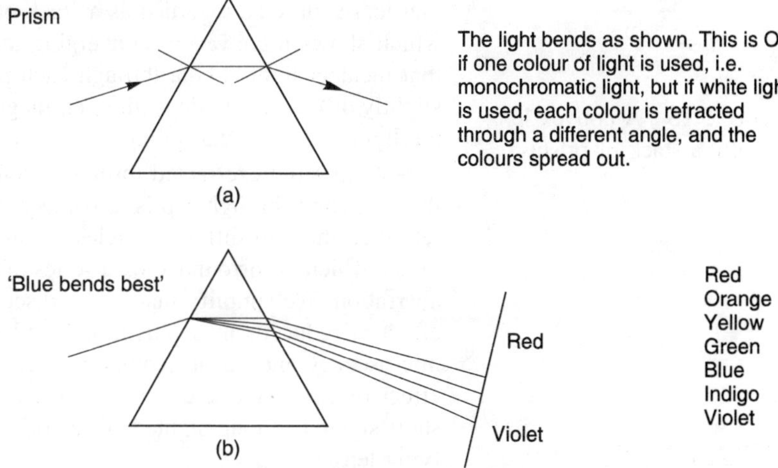

Figure 5.12 (a) Monochromatic light bending through a prism. (b) White light bending through a prism

The prism

Light incident on a prism bends as shown in Figure 5.12(a). This is fine if only one colour of light – called monochromatic light – is used, but if white light is used, each colour is refracted through a different angle and the colours spread out as shown in Figure 5.12(b).

Pure white light is made up of all the colours of the rainbow – red, orange, yellow, green, blue, indigo and violet. If necessary this can be remembered from the quotation 'Richard of York gave battle in vain', since the initial letters of each word correspond to the colours of the rainbow. Frankly, I've never seen the need for this since it's just as easy to commit the order of the colours themselves to memory as it is to recall

Richard's historical activities, doomed to failure as they were! As far as the actual bending of different colours of light is concerned, however, I've found it useful to remember 'blue bends best' (even though it's not true), because it's a good reminder that it is the blue end of the spectrum which is refracted the most.

In a colour television, it has been found that combining just green, blue and red light will give a good approximation to white light, so only those three are used. However, while we are talking prisms, let's move onto the action of lenses and return to colour televisions later.

Lenses

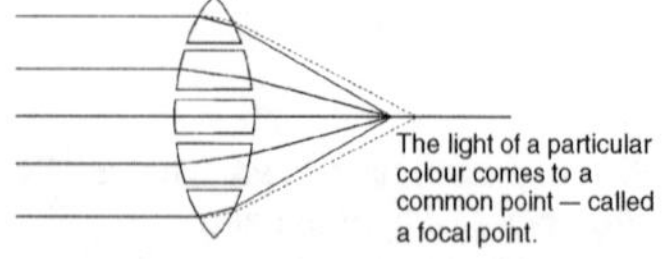

The light of a particular colour comes to a common point — called a focal point.

Figure 5.13 A convex lens is really a stack of prisms

Lenses are very familiar objects, as spectacle wearers and camera owners will know. The subject can become very complicated when studied in detail but we really only need to know a couple of things. The first is that lenses may be regarded as a stack of prisms as shown in Figure 5.13 which shows a convex or converging lens. The second thing to know is that incident light is bent through each prism (or section of the lens) in a slightly different way depending on its geometry, the result being to focus the light at a particular point.

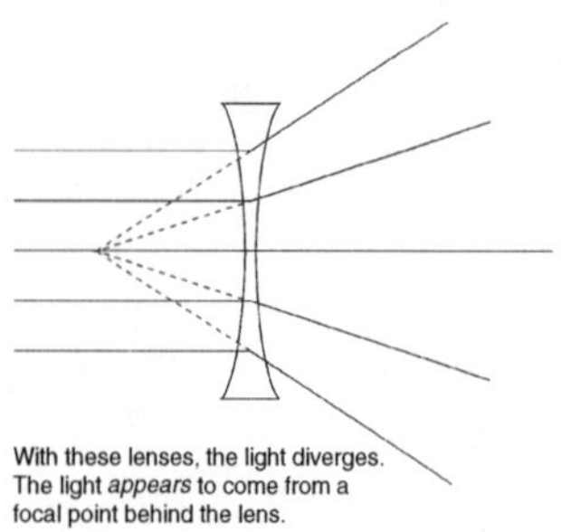

With these lenses, the light diverges.
The light *appears* to come from a focal point behind the lens.

Figure 5.14 The action of a diverging lens

Note that light refracted through a lens is subject to the same rules as that refracted through a prism, namely, that different colours of light are refracted through different angles. A simple lens, therefore, produces an image which is surrounded by a series of coloured bands; this is chromatic aberration. We simplify matters by discussing only thin lenses, where the effect is reduced. In sophisticated optical instruments like cameras and microscopes and so on, compound lenses are used which eliminate the effect of chromatic aberration. You can always tell whether a person is short-sighted or long-sighted since correction of the latter requires magnifying lenses.

Concave, or diverging, lenses cause incident light to be diverged. The light appears to come from a focal point behind the lens as shown in Figure 5.14.

Colour mixing with light

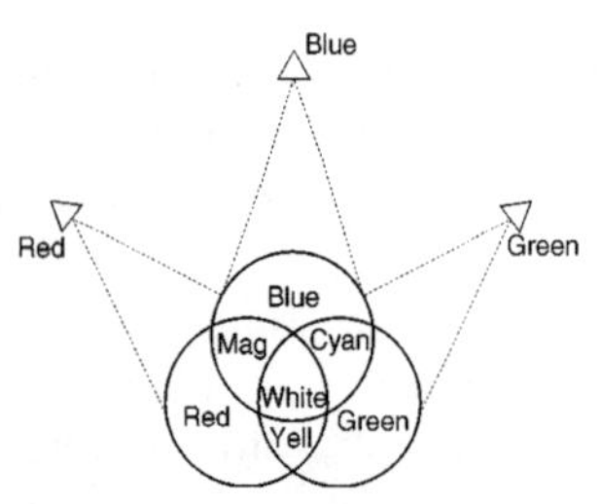

Figure 5.15 Colour mixing with light

Three overlapping lights – red, green and blue, called primary colours – of equal power produce the main colours which we see in everyday life. A combination of red and blue produces a purple-like colour called magenta. Red and green together produce yellow, and blue and green produce (not surprisingly) a bluey-green called cyan. Red, blue and green together produce white light. All of this is shown in Figure 5.15.

For test purposes, colour bars are often shown on the screen of a colour television. They consist of eight bars along the screen starting with white and ending with black and containing the main colours: yellow, cyan, green, magenta, red and blue. In order to remember these colours, Richard of York doesn't get a look in, instead it has been suggested that you

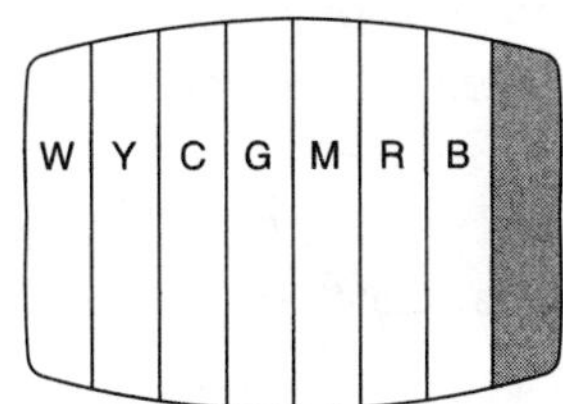

Figure 5.16 The colour bars as they appear on a colour TV screen

remember 'When your cat goes mad, ring Basil Brush'. The origin of this bit of whimsy is unknown, but it certainly sticks in the mind.

Objects appear to be a certain colour because of the light they reflect. An object looks red, for example, because it reflects red light and absorbs all the others. So, an object which is black absorbs all colours and reflects none; a white object reflects all the colours and absorbs none. If a blue sheet of paper is illuminated with monochromatic red light, the blue sheet would look black. Similarly, a yellow sheet would look black if illuminated by monochromatic blue light. This helps to explain why cars of a certain colour look strange when illuminated by yellow, sodium street lights.

Questions

1 State the three ways in which heat is conveyed.

2 Describe methods of heat dissipation for the following:

(a) Soldering a temperature sensitive device.
(b) Protecting a power transistor during operation.
(c) Increasing the cooling effect inside a computer.
(d) Provision of a heat sink for an IC amplifier such as the LM380.

3 Describe the effect of an increase in heat on the following:

(a) A wirewound resistor.
(b) A copper connecting wire.
(c) A silicon transistor.
(d) A PTC thermistor.
(e) An NTC thermistor.

4 State the units of: (a) mass, (b) force, (c) work.

5 (a) What force acts on a mass of 10 kg due to gravity? (Assume that $g = 10$ N/kg.)
(b) A mass of 15 kg is lifted through a vertical height of 10 m. Calculate the work done.

6 An electric motor lifts a mass of 5 kg through 2.5 m in 25 s. What power is developed by the motor?

7 Which of the units, kg, newton or joules, should be used for the following?

(a) mass
(b) weight
(c) energy
(d) work done
(e) kWh
(f) force
(g) force $\times$ distance.

8 An aluminium coil has a resistance of 19 Ω at 0°C. Calculate its resistance at 95°C if its temperature coefficient of resistance is $4 \times 10^{-3}\,K^{-1}$.

9 The distance between peaks of waves on a pond is 0.5 m. Twenty waves reach the bank in four seconds. Calculate: (a) the period, (b) the speed of the waves.

10 A BBC radio programme transmits on a wavelength of 1500 m. The speed of radio waves is 300 000 km/s. Calculate: (a) the frequency, (b) the period.

11 Certain earthquake waves have a period of 0.25 s. They travel at 8 km/s. Calculate their wavelength.

12 A tuning fork vibrates once in 4 ms. The wavelength of the sound produced in air is 1320 mm. Calculate their speed.

13 A signal generator produces a frequency of 3×10^{10} Hz. A transducer 'injects' the waveform into a vacuum tube 5 cm long. How many complete waves will there be in the length of the tube?

14 Certain UHF television waves have a frequency of 625 MHz. Calculate the wavelength. Compare your answer with the approximate length of the rods on a TV aerial given that they will be about a quarter of the wavelength.

15 Calculate the frequency of a mercury vapour lamp producing blue light with a wavelength of 436 nm.

16 A ray of light is reflected from a mirror. If the angle of incidence is 30° what is the angle of reflection?

17 A beam of white light enters one side of a prism. A coloured image containing all the colours of the rainbow appears on a white card placed at the other side of the prism. (a) Which colour is bent the least, and (b) which colour is bent the most?

18 A TV test signal consists of eight vertical colour bars.

(a) State the colours of the bars.
(b) Which of the bars are primary colours?
(c) What combination of coloured light produces the other colours?

19 A yellow card is illuminated with monochromatic red light. What colour will the card appear to be?

20 Why do some lenses show chromatic aberration?

Multiple choice questions

1 Heat transfer by convection can take place in

A solids and liquids
B solids and gases

C liquids and gases
D only in liquids

2 Work done is measured in

A joules
B newtons
C ergs
D dynes

3 The weight, in newtons, of a 10 kg mass is approximately

A 1 N
B 10 N
C 100 N
D 1000 N

4 Heat sinks should ideally

A be painted white
B be made from plastic
C have a large surface area
D be earthed

5 The unit of mass in the SI system of units is the
A newton
B kilogram
C joule
D cubic metre

6 The frequency of a man's voice is 165 Hz. The velocity of sound in air is 330 m/s. The wavelength, in metres, is

A 1 m
B 2 m
C 4 m
D 8 m

7 Yellow light has a wavelength of 600 nm. The frequency is

A 3×10^{14} Hz
B 5×10^{14} Hz
C 5×10^{17} Hz
D 15×10^{17} Hz

8 Mixing red and green light produces

A yellow light
B magenta light
C orange light
D cyan light

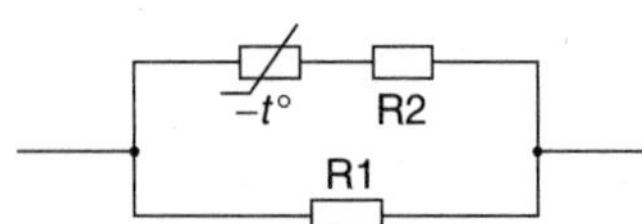

Figure 5.17

9 Manganin is much more suitable than copper for making a wirewound resistance because

A it has a much lower temperature coefficient of resistance
B it has a much higher temperature coefficient of resistance
C it has a much lower resistance
D it has a much higher resistance

10 A negative temperature coefficient thermistor has

A a low resistance at low temperature
B a high resistance at low temperature
C a high resistance at high temperature
D a temperature independent function

11 In Figure 5.17, if the thermistor is heated externally, the total effective resistance of the network will

A always be less than the value of R1
B rise above the value of R1
C be equal to R1 at some temperature
D start above R1 and then reduce

12 Figure 5.18(a) shows a simple form of electronic thermometer. Figure 5.18(b) shows a form of temperature gauge for use on a car engine. The thermistors shown in Figures 5.18(a) and 5.18(b) respectively will be

A both NTC types
B both PTC types
C NTC in Figure 5.18(a) and PTC in Figure 5.18(b)
D PTC in Figure 5.18(a) and NTC in Figure 5.18(b)

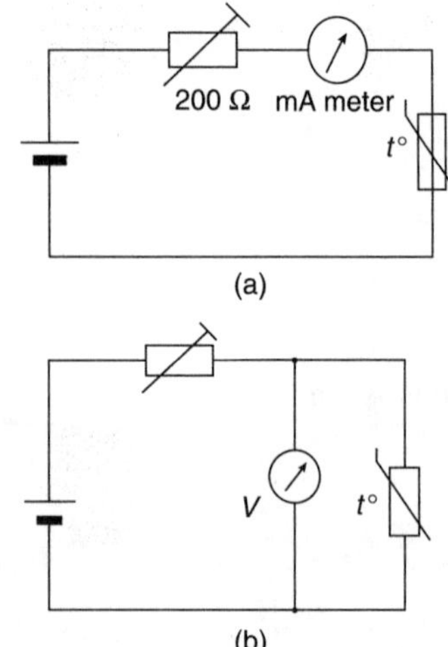

Figure 5.18

Part Two
Systems

6 Power supplies and the tape recorder

Power supplies

We looked at the transformer in Chapter 4 and return to it now for a common, practical application – the power supply unit or PSU.

Say we wish to obtain a 12 V DC supply (the same as is provided by a car battery) from the 240 V AC mains. The first step is to transform the 240 V mains down to *about* 12 V. The second step is to use one or more diodes to change the AC to DC, or, more correctly, to remove the negative-going parts of the waveform. This is shown in Figure 6.1.

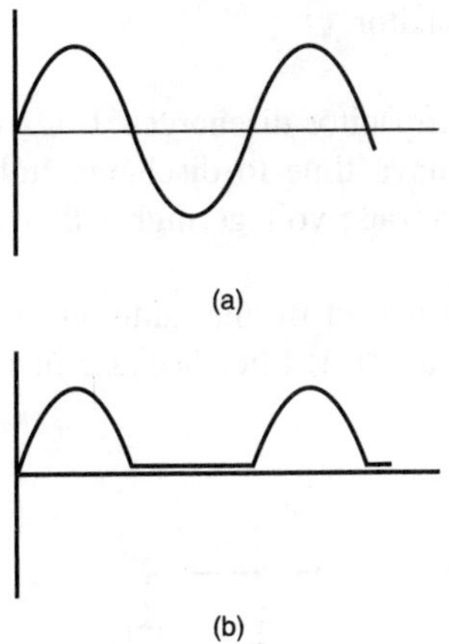

Figure 6.1 (a) The 12 V Sine wave. (b) The rectified waveform with the negative going parts of the waveform removed

A simple method of rectification is shown in Figure 6.2. The diode will only conduct on the positive-going parts of the cycle, the negative-going parts don't get through the diode.

Because only half of the waveform appears across the load, R_L, this method of rectification is called half wave (HW) rectification. It is not usually used in electronics because if half the cycle is lost then so too is half the power. The supply is said to contain a lot of ripple. Figure 6.3 shows the output from a 12 V car battery compared to the output of an HW rectified power supply.

Clearly this is not really very good, because only the peaks of the waveform are at 12 V, the rest of the time it is less than 12 V and at some points is almost zero. This situation can be improved by adding a capacitor which has the effect of smoothing out the ripple. This is shown in Figure 6.4.

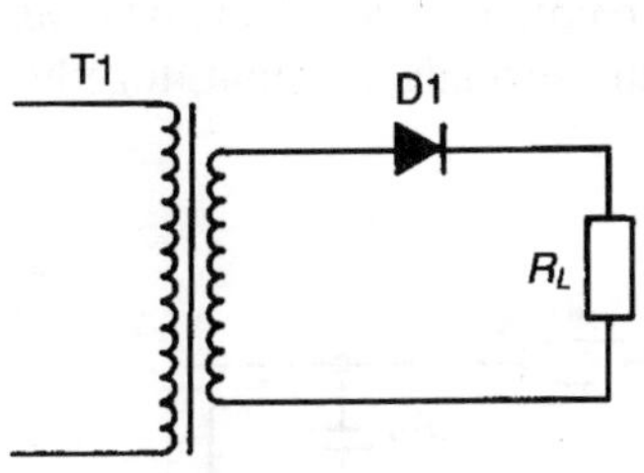

Figure 6.2 The transformer, T1, the rectifying diode, D1, which allows only the positive-going parts of the AC cycle to pass through it, R_L is the load

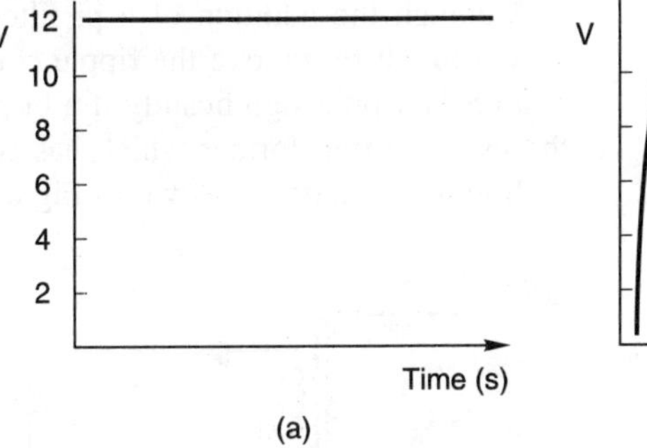
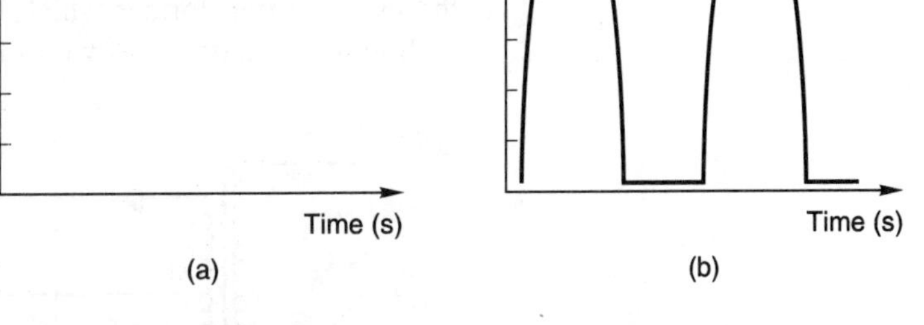

Figure 6.3 (a) Smooth DC supply from a car battery. (b) The output from a half wave rectifier

In Figure 6.4(b), the graph shows the rise to the peak voltage at switch-on. At the same time, the capacitor C1 is charged up to the peak voltage.

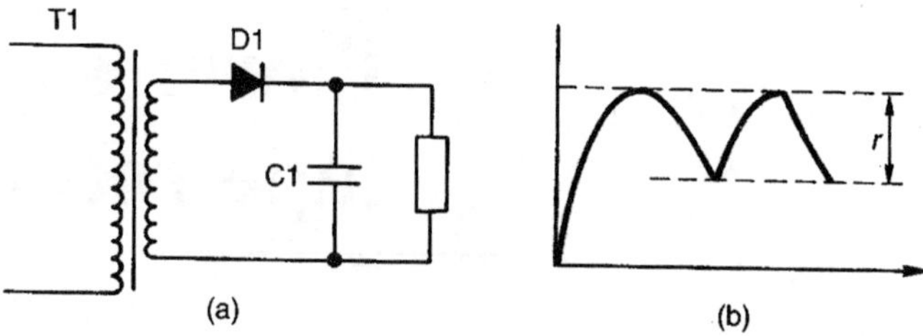

Figure 6.4 (a) The simple HW rectifier with a capacitor, C, added. In (b) the graph shows the rise to the peak voltage at switch-on and the effect on the waveform of adding a smoothing capacitor

As the waveform reduces towards zero, the capacitor discharges. If C1 is made large enough (say 1000 μF) it won't have time to discharge fully before the next cycle appears, keeping the average voltage higher than it would be without the capacitor.

The capacitor smoothing effect can be improved by the addition of a resistor and second capacitor. This is called a π (pi) filter because of its shape, as shown in Figure 6.5.

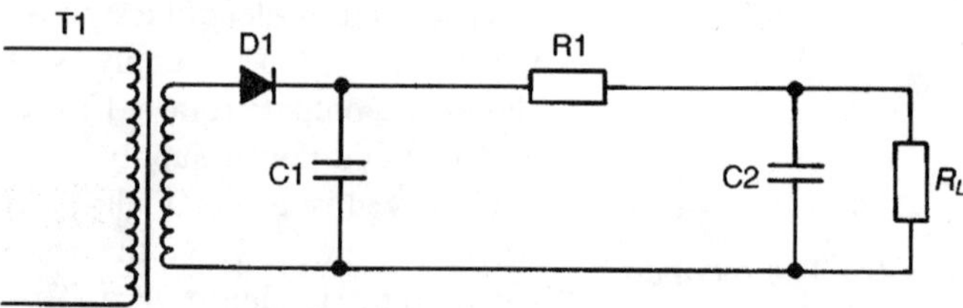

Figure 6.5 The use of a pi filter for increased smoothing. C1 is called the reservoir capacitor, C2 the smoothing capacitor

Although the addition of a pi filter improves the output, it is always very difficult to remove the ripple completely. However, it is possible to reduce the ripple significantly if a bi-phase rectifier is used. This involves the use of a transformer which has twice the secondary output, together with another diode as shown in Figure 6.6.

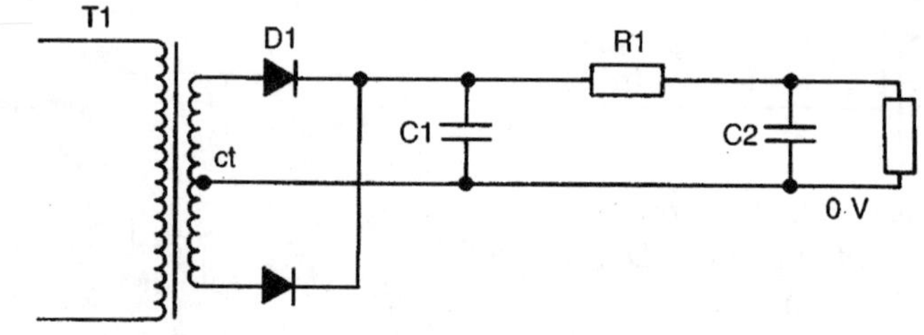

Figure 6.6 A bi-phase rectifier circuit. The point marked ct is the centre tap which becomes the 0 V line

The main problem with the bi-phase rectifier is the increased size, weight and cost of the transformer required. A neat solution to the problem is to use a bridge rectifier as shown in Figure 6.7.

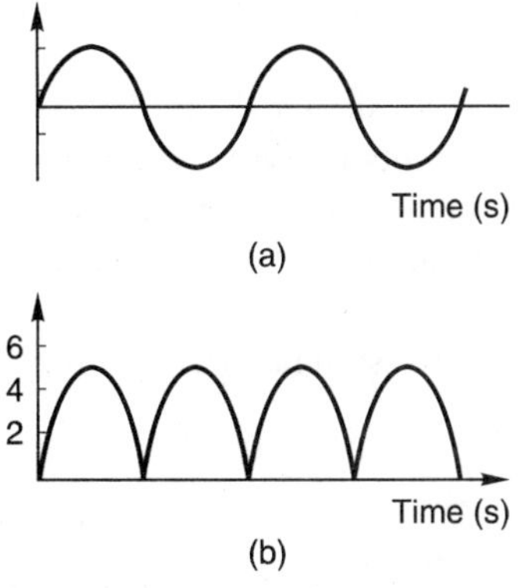

Figure 6.8 (a) The sine wave output from the transformer. (b) The full wave output from the bridge rectifier. Both the bi-phase and bridge rectifiers are full wave (FW) rectifiers with a ripple of 100 Hz

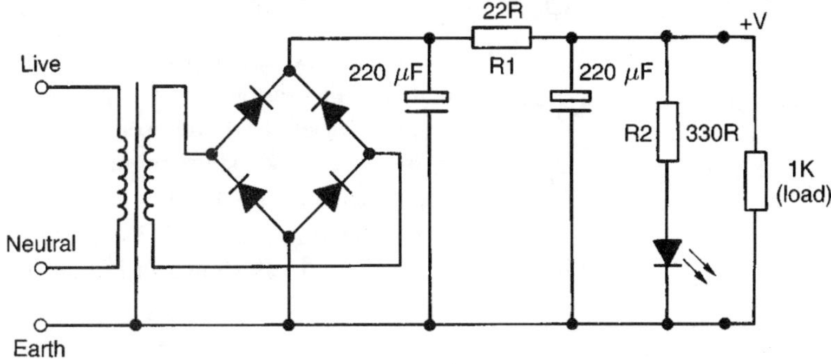

Figure 6.7 The use of a bridge rectifier

The use of a bi-phase or bridge rectifier produces full wave (FW) rectification which effectively increases the ripple frequency from 50 Hz to 100 Hz and the ripple at 100 Hz is much easier to remove. The waveforms are shown in Figure 6.8.

Although most of the ripple may be removed using a full wave rectifier and pi filter, there is still the problem of regulation. This refers to the change in voltage at the output of the PSU due to an increase in load current, which may be simulated by using a *lower* value load resistor causing more current to flow. An increase in current will produce a larger voltage drop across the secondary of the transformer (usually negligible), a greater voltage drop across the resistor in the pi filter and, since the load value is reduced, a smaller path for the smoothing capacitor to discharge through.

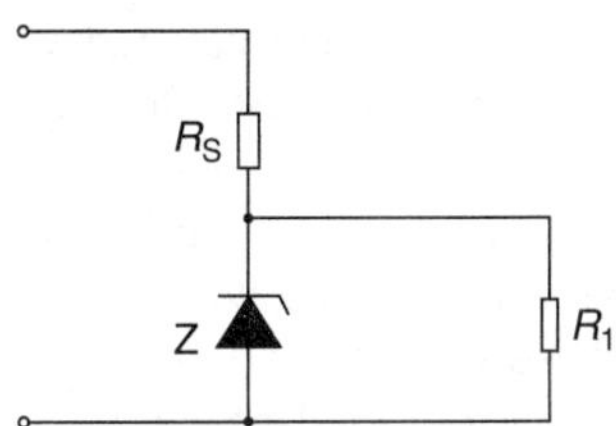

Figure 6.9 A simple shunt regulator where a zener diode is placed in parallel with the load

The solution is to use a regulator or stabiliser which will maintain the output voltage at a constant value even when load conditions vary. This requires a larger voltage on the secondary of the transformer and the addition of the stabiliser components. A simple 'shunt' regulator is given in Figure 6.9.

The zener diode

The ordinary diodes used in the rectifying circuits just described only conduct in the forward direction, i.e. whenever the anode is more positive than the cathode. If an attempt is made to force more current through the diode in the opposite direction (known as reverse biasing) a point will be reached where electrical breakdown occurs with the consequent destruction of the diode. Zener diodes do not do this and so they can be used in a simple regulator circuit.

You will note that the zener in Figure 6.9 has its cathode connected to the more positive side of the supply and is therefore reverse biased.

Provided a certain minimum current flows (say about 10 mA) the diode conducts in the reverse direction at a given voltage (say 9 V).

If the load current decreases then more current flows through the zener diode, but it still has 9 V across it. In this way, the zener diode stabilises the PSU output to 9 V. Figure 6.10 illustrates two common types of zener diode.

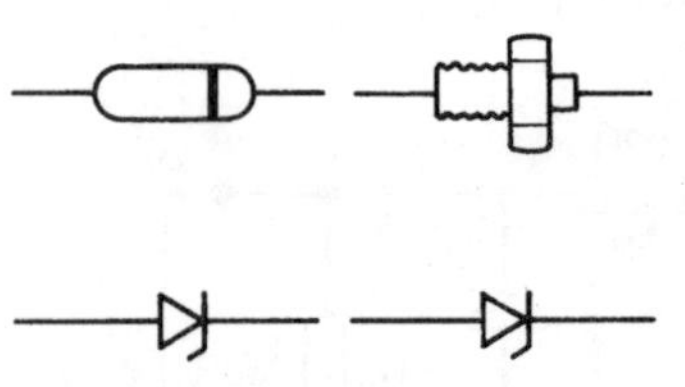

Figure 6.10 Some common types of zener diode

At least two symbols are in common use, as shown. Note, however, that the illustrations are not paired with the symbols. Either symbol can be used for either diode. You will not be required to describe the working of the zener diode shunt regulator, or calculate the values of the associated components at this level.

All the stages in a PSU may be conveniently summarised by using a block diagram as shown in Figure 6.11.

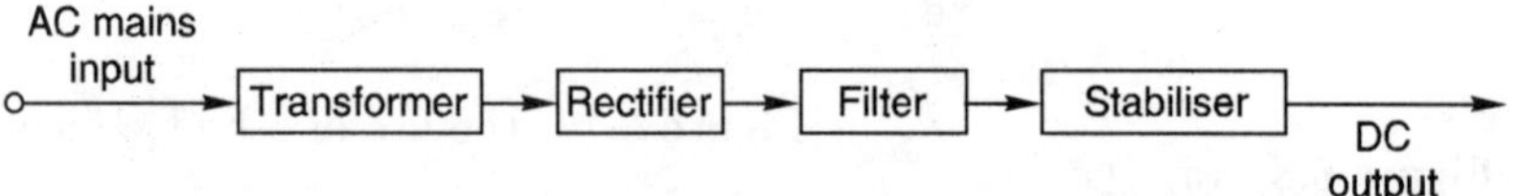

Figure 6.11 Complete block diagram of power supply unit

The tape recorder

All modern tape recorders use ring-shaped electromagnets in order to produce a magnetic field that is constantly varying in sympathy with an audio signal fed to it as shown in Figure 6.12.

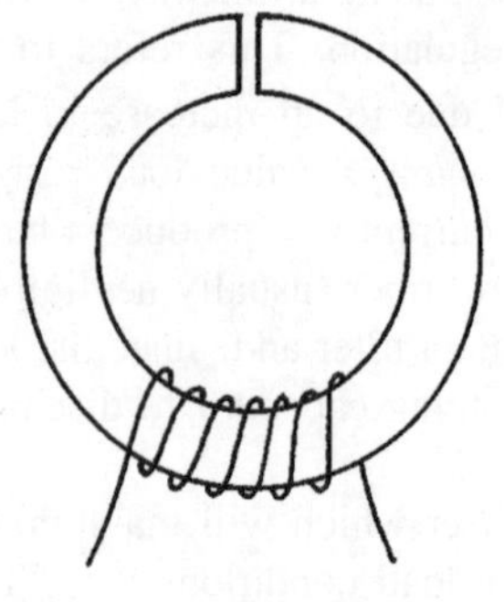

Figure 6.12 Ring-shaped electromagnetic used as a tape head

If a magnetisable substance is placed at the gap, then it can become magnetised in the same way. In a tape recorder, thin tape covered with small magnetic particles – often iron oxide – is pulled across the gap in the recording head. As the tape passes, the particles are aligned in a way determined by the magnetic field produced by the head. A similar head is used for playback, where the tape is drawn across it and which detects the varying magnetic patterns on the tape. These small variations produce a signal voltage which is then amplified and fed to a loudspeaker.

Although this sounds quite simple in theory, there is one important modification that must be included in order to make the system operate satisfactorily. The relationship between a magnetic field and the magnetising field that produced it is not a linear one.

What this means in practice is that any system of recording using the magnetic techniques described would not produce an undistorted output. However, the application of *bias* to the recording signal will result in its being *shifted* into the linear region. The signal that produces this effect is called the *bias* signal and is usually a sine wave with a frequency of between about 40 kHz and 100 kHz. The choice of these frequencies ensures that the bias signal cannot be heard.

The bias signal is mixed with the audio signal being recorded and the two signals are then taken to the record head. The bias signal also doubles as an erase signal. A head, very similar to the record head, is positioned such that as the tape travels past it in record mode, the bias signal erases any previous recording. The record head is positioned just a centimetre or two from the erase head enabling the new signal to be recorded.

On more sophisticated machines, there may be separate record and playback heads. However, we assume a two-head machine when constructing a block diagram for the complete system shown in Figure 6.13.

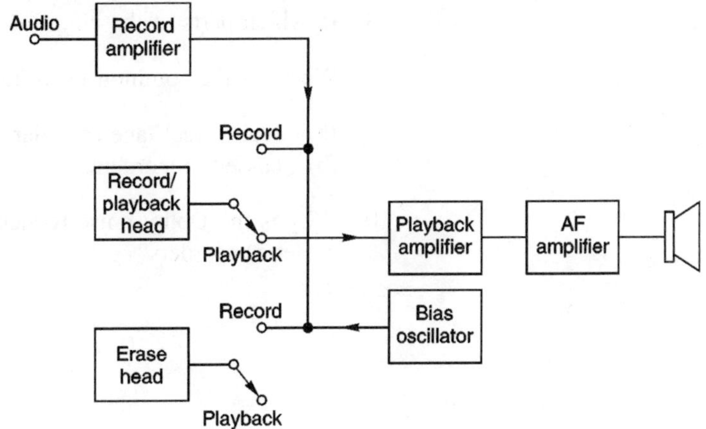

Figure 6.13 Complete block diagram of tape recorder

The Dolby system of noise reduction

This is a system for reducing tape hiss. When reproducing the loudest sounds without distortion, the softest sounds are of such low intensity that electronic noises become noticeable in the background. In the Dolby system an electronic circuit is used which can vary the treble equalisation curve so that during loud passages the boost is applied. No tape overloading then occurs since the highest boost levels are only used during the quietest parts of the recording. During replay, a decoder is used to achieve the correct frequency response. The maximum treble cut occurs during the replaying of quiet passages reducing noise and hiss.

The system allows almost noiseless reproduction and, since it is most effective at slower tape speeds, is of particular value in cassette recorders which run at $1\frac{3}{4}''$ per second. Reel-to-reel tape recorders generally run at $3\frac{1}{2}''$ per second while for superior recording and playback quality, a speed of $7''$ per second is used.

Questions

1 Draw the BS symbol for a diode and label the anode and cathode.

2 Draw the circuit diagram of a simple half wave rectifier without any smoothing components.

3 State a typical value of smoothing capacitor in a simple power supply circuit.

4 Draw a diagram of bi-phase rectifying circuit complete with pi filter.

5 What is the name of the extra circuitry which keeps the output constant even when the load current varies?

6 State a typical tape recorder bias oscillator frequency.

7 Why is a high frequency required for the bias oscillator?

8 In which parts of the circuit is the bias oscillator used?

9 What are the common recording speeds for:

(a) reel-to-reel tape recorders
(b) cassette recorders.

10 Why is the Dolby noise reduction system especially useful in audio cassette recorders?

7 AM and FM transmission and reception

Radio waves

Radio waves are electromagnetic waves (em waves) which consist of varying electric and magnetic waves at right angles to each other and the direction of propagation. Em waves travel at the speed of light in a vacuum. Their velocity is reduced in more dense media; this is the opposite to sound waves which travel faster in more dense media and will not propagate at all in a vacuum. The wave equation, however, applies to all waveforms: $v = f\lambda$.

We have already seen that the wavelength of a waveform is the length of one complete cycle. Its frequency is the number of waves in one second. For all em waves in a vacuum (and to a very close approximation in air) their velocity is equal to c, which is $3 \times 10^8\,\mathrm{ms}^{-1}$. They can exist at many different frequencies forming the electromagnetic spectrum and these are shown in Table 7.1.

When we speak the air around us is being constantly alternately compressed and rarefied. These vibrations are termed mechanical waves; our ears can detect them while our brain decodes the information so that we can recognise the spoken word. A man's voice ranges in frequency from about 250 Hz to 2500 Hz. Our ears will respond to frequencies from about 20 Hz up to 20 kHz although this varies from one person to another and is usually age dependent, older people, for example, generally having a reduced range.

If sound waves were converted into radio waves of the same frequency (which is quite possible) we wouldn't be able to hear them as our ears only respond to mechanical vibrations. Choosing such a low frequency, however, would limit sound broadcasting to only one station. Furthermore, when RF signals are transmitted, the aerial or antenna being used must be a certain length, associated with the wavelength of the transmitted signal. Let's say we decide that an 'average' frequency for the spoken voice is 1 kHz, then the wavelength using the wave equation, $v = f\lambda$, would be:

$$\lambda = v/f = 3 \times 10^8/1000 = 300\,000\,\mathrm{m}!$$

For effective transmission, a radio antenna has to be a minimum length of 1/8 of the transmitted wavelength and even this would amount to the antenna being 37 500 metres long! So, higher frequency radio waves are used:

(a) so that more than one station can be transmitted simultaneously
(b) so that the transmitting antenna and receiving aerial can be a sensible length (generally 1 m or less).

Table 7.1 Frequencies in the electromagnetic spectrum

	Radio waves	*IR*	*Visible light*	*UV*	*X-ray*	*γ-ray*	*Cosmic*
F	30 kHz–960 MHz	10^{12}–10^{14}	10^{14}–10^{15}	10^{15}–10^{17}	10^{16}–10^{20}	10^{19}–10^{21}	10^{21} Hz
λ	10^{4}–10^{-2}	10^{-3}–10^{-5}	10^{-6}–10^{-7}	10^{-7}–10^{-9}	10^{-10}	10^{-12}	10^{-14} m

Note that the convention of using an 'antenna' for radio transmission and an 'aerial' for radio reception has been adopted here. There are those who argue about this. Generally the two terms are interchangeable without problems.

Modulation

Radio transmission first of all involves the generation of a high frequency signal. This is called the carrier wave. The signal to be transmitted, the spoken voice or music, for example, is called the modulating signal. The process by which the carrier wave is modified by mixing it with the modulating frequency is called modulation. In most cases, the carrier frequency is much higher than the modulating frequency, the carrier being a radio frequency and the modulating frequency being an audio wave.

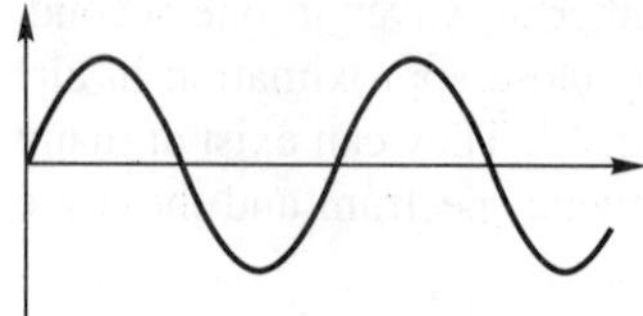

Figure 7.1 The modulating signal – a simple sine wave

Since the sounds of the human voice involve a large number of frequencies produced simultaneously, the form of the modulating wave is rarely sinusoidal. However, to make the analysis simple, we will assume that the modulating wave is sinusoidal as shown in Figure 7.1.

Amplitude modulation

It will be shown that both the process of modulation and detection (the recovery of the modulating signal at the receiver) involve the generation of frequencies not originally present. Put simply, the extent by which the RF carrier is modulated depends upon the modulation factor, M, of the modulating waveform. The modulation factor is also called the degree of modulation and this depends upon:

(a) the amplitude of the modulating waveform
(b) the amplitude of the RF carrier wave
(c) the characteristics of the modulating circuit.

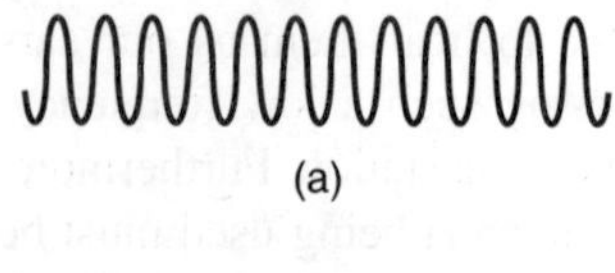

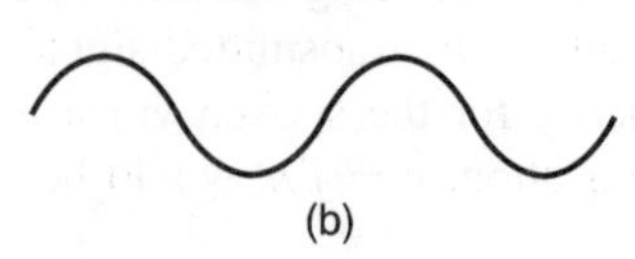

Figure 7.2 Waveforms in AM transmission. (a) The unmodulated carrier. (b) The modulation waveform

There is no direct relationship between the envelope amplitude and its frequency. In amplitude modulation, the frequency of the carrier wave is constant but its amplitude is varied at the modulation frequency. Figure 7.2(a) shows an unmodulated carrier wave; the modulation waveform (the audio signal required to be transmitted) is shown in Figure 7.2(b).

When the carrier and audio waveforms are combined (by the modulating process), the waveform looks like those shown in Figures 7.3(a) and 7.3(b).

When M lies between zero and unity, the modulated wave is of the form of the curve shown in Figure 7.3(a). When M is unity, the amplitude varies

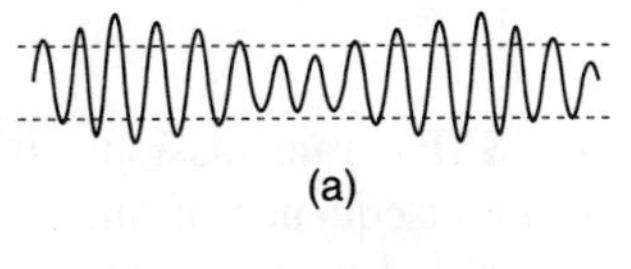

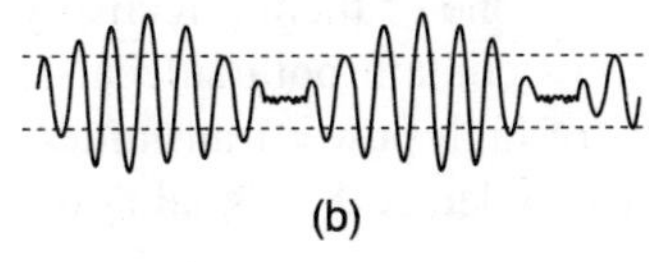

Figure 7.3 (a) Modulation when $M = 0.5$. (b) Modulation when $M = 1.0$, when modulation is said to be complete

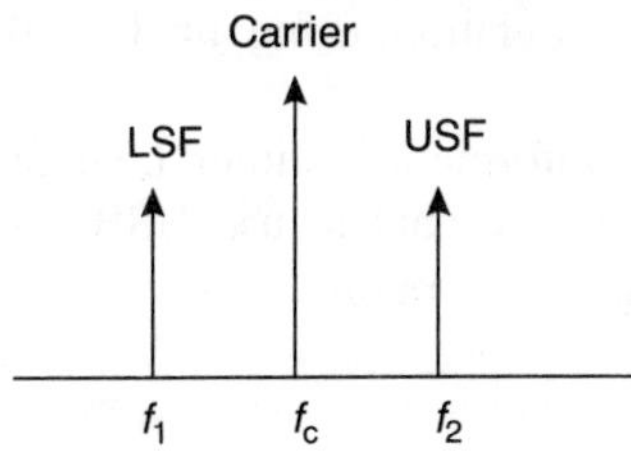

Figure 7.4 Side frequencies produced by a signal sine wave

between zero and twice the unmodulated value as shown in Figure 7.3(b), where modulation is said to be complete.

Side frequencies in AM

An amplitude-modulated wave is made up of three components having frequencies equal to the carrier frequency, the carrier frequency plus the modulation frequency, and the carrier frequency minus the modulation frequency. The process of amplitude modulation therefore involves the generation of sum and difference frequencies, called the upper side frequency (USF) and the lower side frequency (LSF) respectively. These are usually shown diagrammatically as in Figure 7.4.

If f_c is the carrier frequency (100 kHz in this case) and f_s is the signal frequency providing the modulation (1 kHz in this case), then:

$$f_1 = f_c - f_s \quad (= 99\,\text{kHz})$$

$$f_2 = f_c + f_s \quad (= 101\,\text{kHz})$$

Remember, for simplicity only one signal frequency has been used so that the carrier is modulated by a pure sine wave. In normal broadcasting, of course, this is not the case; the transmission of music or speech involves many of the audio frequencies. These range between about 20 Hz and 20 kHz, so that instead of producing single side frequencies, a band of frequencies is produced and these are called sidebands. A band contains a number of different frequencies and the bandwidth is the *range* of frequencies occupied by a signal. Since sound radio broadcasts consist of audio frequencies, then the bandwidth would be about 20 kHz either side of the carrier frequency. This is shown diagrammatically in Figure 7.5.

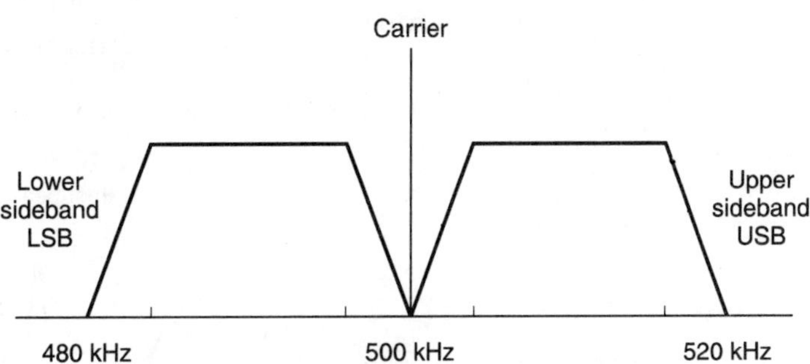

Figure 7.5 Sidebands produced when the modulating signal covers the audio range (about 20 Hz to 20 kHz)

For viable transmission, the bandwidth required must be large enough to contain all the signal frequencies required to be transmitted. In this case the transmitted bandwidth would be about 40 kHz.

Amateur transmissions

Amateur operators tend to use single sideband (SSB) transmissions to conserve space. The two sidebands produced as a consequence of amplitude modulation are identical so transmitting only one of them effectively halves the bandwidth. Professional transmissions contain both sidebands because there is no shortage of space, the cost of high power transmitters is not much of a problem and it makes it easier to detect the signal at the other end.

Amateurs also limit the bandwidth of their audio (to frequencies between about 30 Hz and 3400 Hz using a filter after the operator's microphone) to reduce the overall bandwidth even further. So, by transmitting only one sideband, amateurs conserve power and all their transmitter power goes into the required signal. As an extension to this, amateurs often suppress the carrier signal as well, since this is only required to produce the sidebands. This is achieved by a special form of modulation, using a balanced modulator.

Since the two sidebands contain identical information, either LSB or USB could be transmitted. However, it is conventional to use USB for frequencies above 10 MHz, and LSB for frequencies below.

Names of frequency bands

Different radio frequencies are allocated to specific uses and each one has a particular name, as shown in Table 7.2.

Modulation depth

This is the ratio of the signal peak to the carrier peak.

$$\text{Modulation depth} = \frac{\text{Signal peak}}{\text{Carrier peak}} \times 100\%$$

These relationships are shown diagrammatically in Figure 7.6.

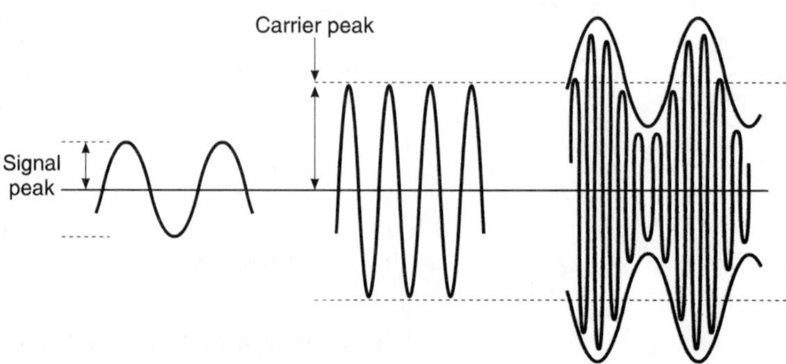

Figure 7.6 Modulation depth

Table 7.2 Names of frequency bands and specimen allocations

Band name	*Frequency band*	*Wavelength range*	*Characteristics and range*	*Service type authorised*
Very low frequency (VLF)	10–30 kHz	30–10 km	Long distance propagation	Standard transmissions
Low frequency (LF)	30–300 kHz	10–1 km	1000 miles by day, less at night	Telegraphy; navigation; sound
Medium frequency (MF)	300–3 MHz	1 km–100 m	200 miles by day, further at night	Sound broadcasting
High frequency (HF)	3–30 MHz	100–10 m	Long distances via ionosphere	Amateur radio operators
Very high frequency (VHF)	30–300 MHz	10–1 m	Visual range	FM sound broadcasting; telephony
Ultra high frequency (UHF)	300 MHz–3 GHz	100–10 cm	Visual range	Terrestrial and satellite television
Super high frequency (SHF)	3–30 GHz	10–1 cm	Microwaves	Radar
			Frequency	*Wavelength*
Extremely high frequency (EHF)			>30 GHz	<1 cm

If, for example, the signal peak is 1 V and the carrier peak is 2 V, then the modulation depth is (1 V/2 V × 100)% = 50%. The value of 50% is also known as the modulation index.

Figure 7.7 might help to clarify the matter if you are still in some doubt.

$$\text{Depth of modulation, } m = \frac{\text{Signal peak}}{\text{Carrier peak}} \times 100\%$$

$$= \frac{\text{Max.} - \text{Min.}}{\text{Max.} + \text{Min.}} \times 100\%$$

AM transmission

Much has been said about modulation and sidebands so let's see how they are produced. The frequency of transmission is dictated by the frequency of the 'carrier' wave; this is a radio frequency (RF) signal usually somewhere

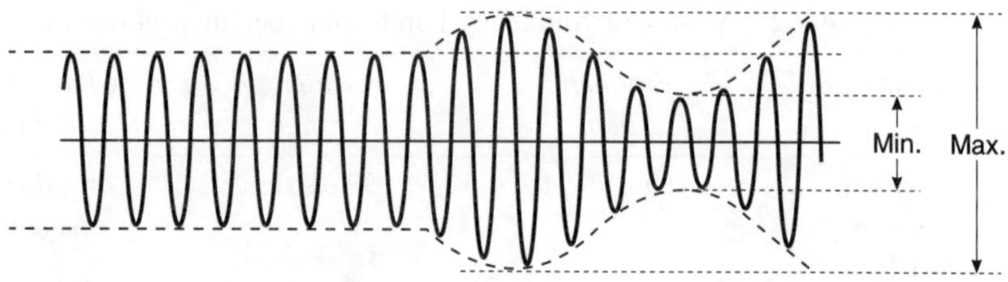

Figure 7.7 A definition of modulation depth

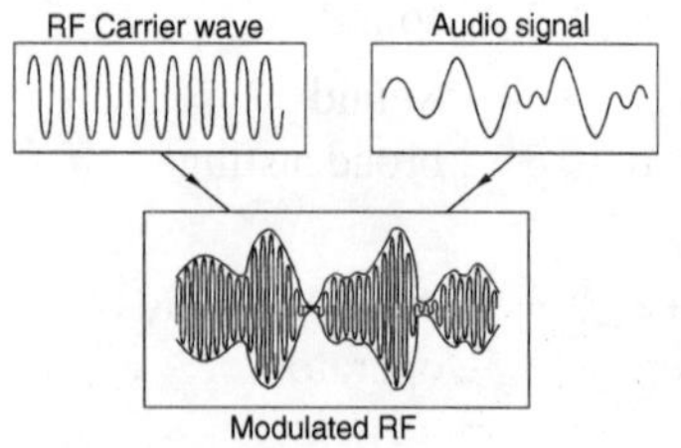

Figure 7.8 Producing amplitude modulation (AM)

between 200 kHz and 2 MHz according to present standards. This signal is produced by a very stable, sine wave oscillator of which there are many kinds; the Colpitts oscillator is one example.

If speech is being transmitted, a microphone is used. This is connected to an audio amplifier which increases the signal to an appropriate level. The constant amplitude carrier wave is then mixed with the audio signal and it is this process which is called modulation. The effect is to change or 'modulate' the RF carrier wave in sympathy with the audio signal being mixed with it. This is shown in Figure 7.8.

The modulated RF contains the carrier frequency and the two sidebands discussed earlier. The only frequency not present is the one that is actually required – the audio signal! However, the audio information is contained in the sidebands and can be recovered very simply by the receiver.

A very simple transmitter, capable of amplitude modulation, is shown in Figure 7.9.

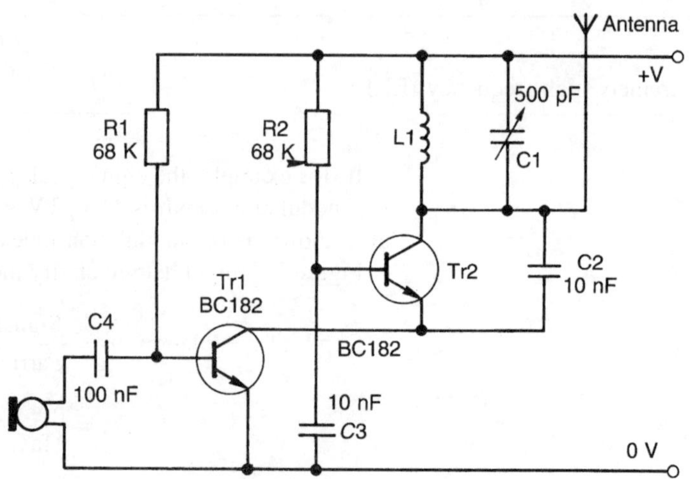

Figure 7.9 A simple AM transmitter

Tr1 forms the emitter load of Tr2 which is itself an RF oscillator. The amplitude of the oscillations is controlled by the emitter load, so while the frequency of the oscillator remains constant, its amplitude is varied if the emitter load is varied. Since, in this case, the emitter load is formed by Tr1, any variations on its base will vary the amplitude of the oscillator waveform. The microphone (probably a high impedance type) is connected via C4 to the base of Tr1 so that it is the variation of the input sound signals which vary the oscillator amplitude. So, simple amplitude modulation is obtained.

Some suggested component values are given; although the circuit can be made to work (which is illegal because radio transmissions of this sort should only be made by licensed operators) the range is very low and construction of this device may therefore be described as a 'laboratory experiment'.

Note The output of the transmitter contains the RF frequency plus the two sidebands. However, harmonics may also be generated so that it would be wise to include a low-pass filter between transmitter output and aerial in order to suppress unwanted, high frequency harmonics. A complete block diagram of an AM transmitter is shown in Figure 7.10.

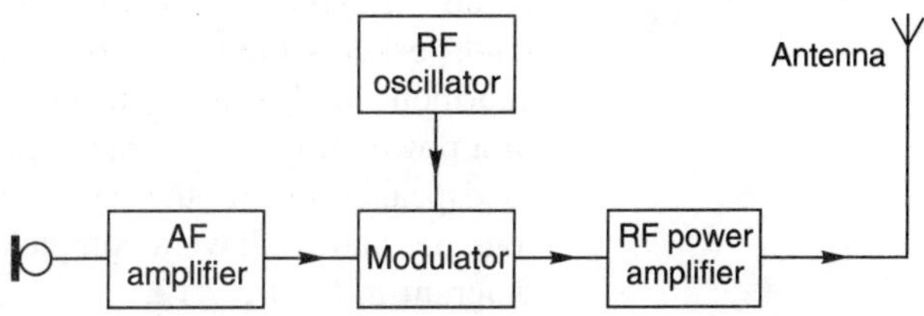

Figure 7.10 Complete block diagram of AM transmitter

AM receiver

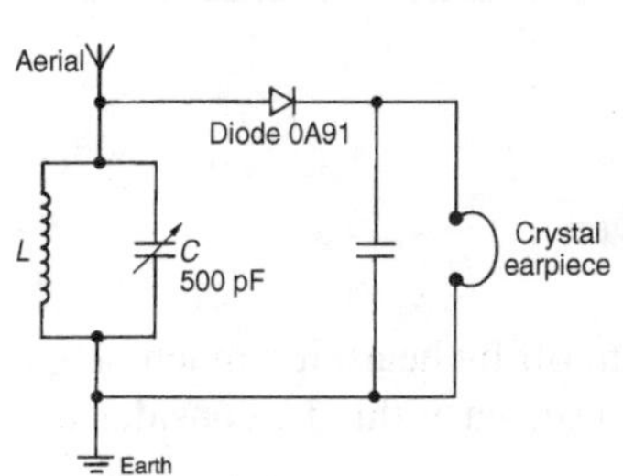

Figure 7.11 Circuit diagram of a simple crystal set

Reception of an AM radio signal can be achieved by the use of a simple 'crystal' set, a circuit for which is shown in Figure 7.11. The 'crystal' of this circuit is the 0A91 point contact diode which is a 'modern' equivalent. The circuit requires no battery to power it, but the signals are very weak and variable.

The coil and variable capacitor form the tuned circuit which allows the operator to receive one signal while rejecting (to some extent) all the others. The diode is the detector (or demodulator, which means the same thing here) and this component removes one of the sidebands. The high impedance earpiece allows the operator to hear the signal.

From this simple circuit, it can be seen that there are three main sections: tuned circuit, demodulator and sound output. All receivers must have these as a basic minimum. Standard receivers, however, have much more; where amplification of the signal is required (as it is in order to drive a loudspeaker) each stage must be capable of being tuned to each programme required. This involves more variable tuned circuits for each stage which introduce complications such as instability, not to mention the cost of so many variable capacitors. In order to eliminate these problems, the 'superhet' receiver was introduced.

Superhet is short for 'supersonic heterodyning'. Briefly, this means combining signals at frequencies higher than those audible to the human ear. The combination, or mixing, produces a beat frequency, lower than the RF signal which is one of its components, but higher than the audio frequency required. Hence, the new frequency is called an intermediate frequency, or IF. This is produced by mixing the incoming RF with an oscillator built into the receiver. As the oscillator's frequency is capable of being varied at the same time as the tuner which selects the incoming signal, the IF frequency remains constant. Any IF stages in the receiver, therefore, are of fixed frequency making their design much less complex and far more stable. In the UK, for AM receivers, the standard IF frequency has been set at 470 kHz.

Fading of the signal, causing changes in volume at the loudspeaker end, may be obviated by using automatic gain control (AGC). This control signal is always taken from the demodulator (or detector) and then fed back to the first IF stage. The standard receiver is completed by the addition of an AF voltage amplifier and the power amplifier which drives the loudspeaker. Note that a power amplifier is *not* necessarily one that drives enormous loudspeakers such as those used in discos; the distinctive quality of a power amplifier is that it can drive a low impedance source such as a loudspeaker, or, in other systems, an electric motor. Put all these sections together and we have a complete AM receiver which is shown as a block diagram in Figure 7.12.

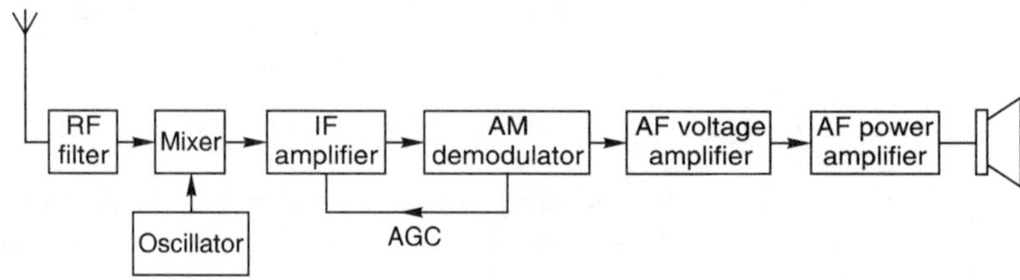

Figure 7.12 Block diagram of AM receiver

Although it is possible to learn the diagram off by heart, it's much better to learn it through understanding. To show how easy this is, consider the following notes:

1 The loudspeaker is driven by the power amplifier (PA).
2 The PA is logically preceded by the AF voltage amplifier.
3 AGC is derived from the demodulator and it is fed back to the IF amplifier.
4 The block on its own is the oscillator.
5 The first block is the tuner, or RF filter as it is designated here, and the next block, which combines the RF filter output with the oscillator, is the mixer.

Remembering the block diagram in this way should enable you to draw it, not simply from memory, but by knowing what each block does and therefore the manner in which each one is logically connected to the other.

The FM receiver

Amplitude modulation (AM) is so called because when modulation takes place it is the amplitude of the carrier which is varied, the frequency of the carrier wave remaining constant. However, when interference occurs, it is normally present on the peaks of the waveform and this can be detected by the receiver. It is passed onto the final stages and hence to the loudspeaker allowing it to be heard. The effect is shown in Figure 7.13.

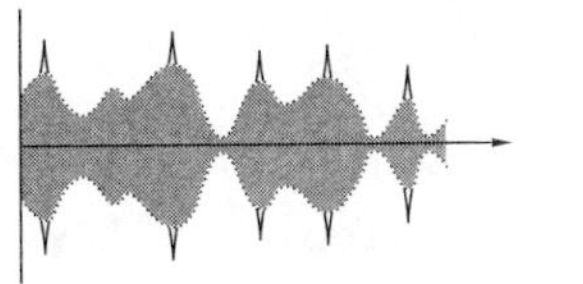

Figure 7.13 Interference spikes on an AM wave

In frequency modulation (FM), the amplitude of the carrier is kept constant and the frequency is varied.

Frequency modulation

Frequency modulation is the process whereby the frequency of one wave or oscillation is varied with time in accordance with the time variations of another wave or oscillation. When the carrier is being modulated, the louder the audio signal, the greater the frequency deviation of the transmitter; the rate at which the frequency changes about its nominal carrier frequency represents the audio frequency transmitted.

Frequency deviation is the change in carrier frequency caused by the modulating signal. For standard FM broadcasts the maximum frequency deviation is 75 kHz and the highest audio frequency transmitted is 15 kHz. This requires a transmission channel of between 165 kHz and 200 kHz.

Many amateurs use FM transmissions but if the deviation were 75 kHz, there would only be room for about ten such transmissions in the 2 MHz bandwidth of the 144 MHz band. For VHF communication, therefore, the deviation is limited to less than ±15 kHz and for amateur radio the signals must not be deviated by more than ±3 kHz so that they occupy about the same bandwidth as an AM signal. Such a system is called narrow-band frequency modulation (NBFM).

FM transmission

This can be effected quite simply by the use of a varicap diode. This is a component whose capacitance is varied by the application of a suitable voltage. If the diode forms part of an oscillator and the voltage on the diode is varied by the application of the audio signal to be transmitted, as the capacitance is thus varied, the oscillator frequency will be varied in sympathy, producing frequency modulation. This is shown in Figure 7.14

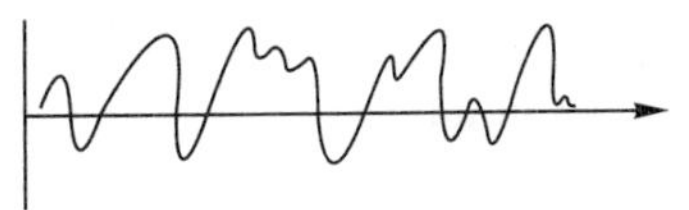

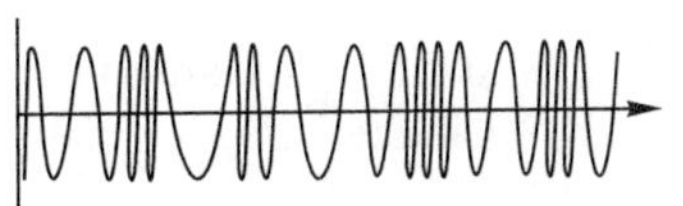

Figure 7.14 An audio modulating wave and its effect on an FM carrier wave

Apart from reduced interference, FM systems also have the advantage of allowing a greater dynamic range of sound intensities to be transmitted. In AM systems, the ratio of the loudest sound to the weakest sound is 200 to 1, while in FM systems, it may exceed 10 000 to 1. FM systems also provide for a very large signal to noise ratio.

AM transmission produces two sidebands as we have seen; in FM transmission, there are (theoretically) an infinite number of sidebands. In practice, however, the amplitude of these sidebands quickly reduces to a negligible value and so can be ignored. Nevertheless, those sidebands which do exist need to be accommodated so that the bandwidth required is much larger than is necessary for AM signals.

Block diagram

A complete block diagram of a typical FM receiver is shown in Figure 7.15.

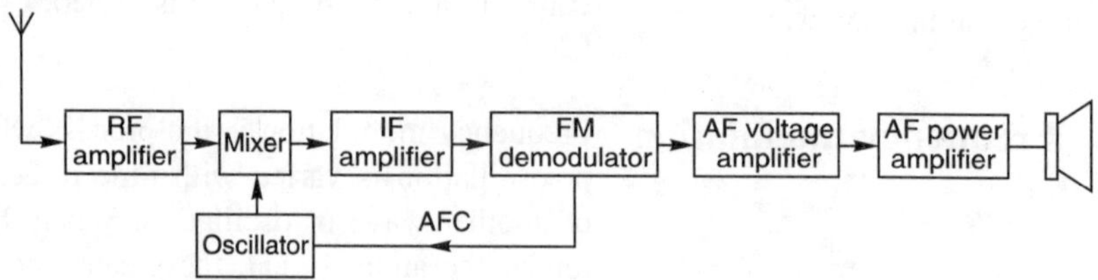

Figure 7.15 Complete block diagram of an FM receiver

Note that this diagram is very similar to the one used for AM reception with the following changes:

1 The RF filter is replaced by an RF amplifier.
2 The AM demodulator becomes an FM demodulator.
3 Automatic frequency control (AFC) is added between the FM demodulator and the local oscillator. This is important because any changes in oscillator frequency are detected as modulation. AGC is not usually present except in very sensitive receivers in order to prevent overload in strong signal areas.
4 The IF frequency is much higher at 10.7 MHz.

Pulse modulation (PM)

There are several different types of PM:

1 Pulse amplitude modulation (PAM)
Here, the modulating wave is caused to amplitude modulate a pulse carrier as shown in Figure 7.16.

2 Pulse time modulation (PTM)
The values of instantaneous samples of the modulating wave are caused to modulate the time of occurrence of some characteristic of a pulse carrier.

3 Pulse duration modulation (PDM)
This is pulse time modulation in which the value of each instantaneous sample of the modulating wave is caused to modulate the duration of

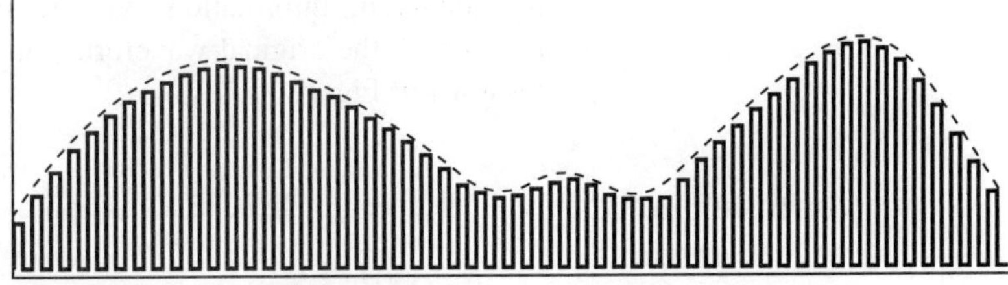

Figure 7.16 Pulse amplitude modulation (PAM)

a pulse. This is also called pulse width modulation (PWM) and pulse length modulation (PLM). PWM is described in more detail later.

4 Pulse position modulation (PPM)
This is pulse time modulation in which the value of each instantaneous sample of a modulating wave is caused to modulate the position in time of a pulse.

5 Pulse frequency modulation (PFM)
This is pulse time modulation in which the pulse repetition rate is varied. This is also called pulse repetition-rate modulation.

6 Pulse code transmission (PCT)
This type of transmission converts the amplitude of each pulse into a form of binary code which is then transmitted.

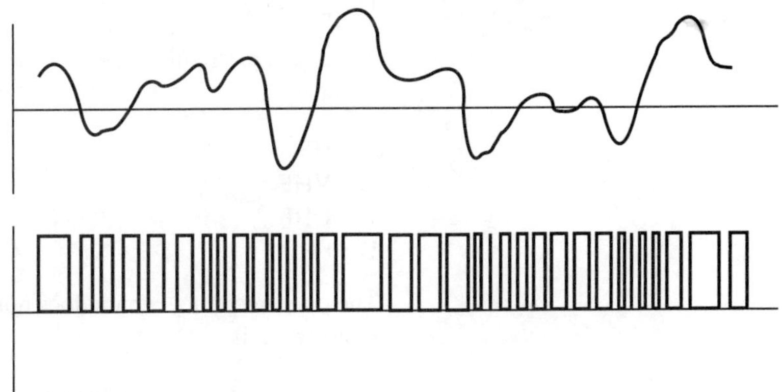

Figure 7.17 Pulse width modulation (PWM)

Pulse width modulation (PWM)

The most common form of pulse modulation is PWM. The great advantage of this form of modulation is that it does not matter how much the shape, amplitude or (to some extent) strength of the signal is varied or distorted, the pulse width can always be regenerated at the receiver. The width of the pulse is related to the modulation which in turn is dependent on the modulating information. As long as the width of each pulse remains undistorted, the original waveform can be recovered in the receiver. This is shown in Figure 7.17.

Multiple choice questions

1 The speed of light in air is approximately

A 330 m/s
B 330 km/s
C 3×10^6 m/s
D 3×10^8 m/s

2 The radio portion of the electromagnetic spectrum is regarded as being approximately

A 20 Hz to 20 kHz
B 20 kHz to 150 kHz
C 30 kHz to 350 MHz
D 30 kHz to 950 MHz

3 A modulated carrier wave has side frequencies of 99 kHz and 101 kHz. The carrier frequency is

A 1 kHz
B 2 kHz
C 100 kHz
D 200 kHz

4 The band of frequencies in which a frequency of 200 MHz would occur is

A HF
B VHF
C UHF
D SHF

5 The only frequency which does not appear in an amplitude modulated radio wave is

A the carrier wave frequency
B the upper sideband frequency
C the lower sideband frequency
D the modulating signal frequency

6 In the UK the standard AM receiver IF frequency is

A 100 kHz
B 470 kHz
C 1 MHz
D 10.7 MHz

7 Standard FM broadcasts have a maximum frequency deviation of

A 10 kHz
B 75 kHz
C 100 kHz
D 1 MHz

8 When applied to radio circuits, the term 'AFC' stands for

A amplifier fine control
B applied frequency conditioning
C automatic frequency control
D adjacent frequency channel

9 Modulation depth is defined as

A $\dfrac{\text{Signal peak}}{\text{Carrier peak}} \times 100\%$
B $\dfrac{\text{Carrier peak}}{\text{Signal peak}} \times 100\%$
C $\dfrac{\text{Bandwidth}}{\text{Audio signal peak}} \times 100\%$
D $\dfrac{\text{Audio signal peak}}{\text{Bandwidth}} \times 100\%$

10 When applied to radio receiver circuits, 'AGC' stands for

A automatic gain control
B amplitude gain control
C amplitude gated channel
D adjacent going channel

8 Speed control system; phase locked loop; the CRT; the CRO; simple colour television

Speed control system

Suppose we have a direct current (DC) motor driving a load where it is absolutely essential that the speed, once set, remains constant and also that the system be resettable at a different speed, but still be held constant at that new speed.

Where constant speed motors are normally used (for example, in record players and clocks) AC motors are used which are governed by the mains 50 Hz supply. These are called synchronous motors because their speed is synchronised with the 50 Hz mains. However, the mains is not always available so that constant speed motors may be of the DC type. The motor is shown as in Figure 8.1.

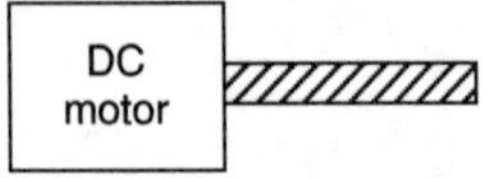

Figure 8.1 A simple representation of a DC motor

The DC motor is not fed from a small DC supply, but instead from a DC power amplifier (PA). This means, among other things, that the output from the DC amplifier can be controlled and varied, thus varying the speed, or, it can be kept very accurately at one level, thus keeping the speed constant. We now need to look at how the output from the DC amplifier is varied and how it can be automatically linked to the speed of the motor.

Speed variation

Since the motor is driven by a DC power amplifier (and not simply from a DC power supply) the speed of the motor can be controlled by the input to the power amplifier. If, as well as driving the load, the DC motor has a small DC generator fitted to it, then the output from the generator (a small DC voltage) will be proportional to the motor's speed. Remember that this small generator only provides a reference signal, it is not large enough to have any effect on the speed of the motor itself, and doesn't tend to slow it down.

The output from the generator is a small DC voltage, proportional to the motor's speed. This voltage is called the feedback voltage because it is fed back to the differential amplifier. The difference between the feedback voltage and the reference voltage is called the error signal. This is passed on eventually to the power amplifier, correcting any change in the speed of the motor and so keeping it constant.

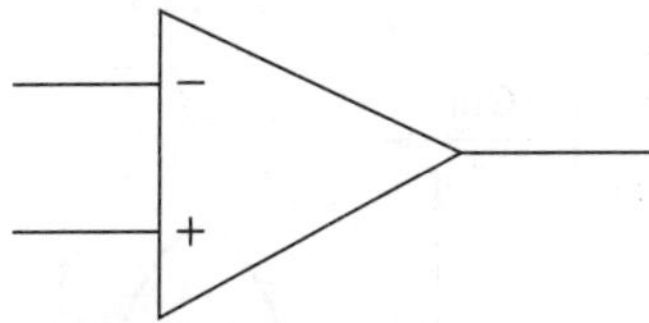

Figure 8.2 A differential amplifier

Differential amplifiers

Figure 8.2 shows a differential amplifier. It has two inputs and the output of the amplifier depends not on the value of the two inputs, but on the difference between them – hence the name, differential amplifier. It can be set to have any gain required, so let's assume it has a gain of 10. This means that the output will be ten times larger than the difference between the two inputs.

If the set-speed voltage is applied to one input of the differential amplifier (say 0.5 V) and the output from the DC generator (say 0.3 V) – the feedback voltage – is fed to the other, then the difference between the two inputs is 0.5 – 0.3 which is 0.2 V. The difference between the two inputs is called the actuating or error signal. If the amplifier has a gain of 10, the output will be 2 V. This is amplified by the power amplifier and then fed to the motor.

If the load on the motor tends to slow it down, the DC generator connected to it must also slow down. If this happens, its output reduces, let's say from 0.3 V to 0.2 V. The difference is now 0.5 – 0.2 which is 0.3 V, and the output becomes 3 V. This is then amplified by the power amplifier, the motor receives a larger voltage and begins to speed up to where it was before. If the motor tends to increase its speed, the feedback signal increases, the difference reduces and so too does the output, and the motor slows down again. In this way, the motor's speed is kept constant.

The set-speed control might simply be a potentiometer across the supply line with limiting resistors at each end. Figure 8.3 shows a complete block diagram of the system.

Figure 8.3 Block diagram of a speed control system

The phase locked loop (PLL)

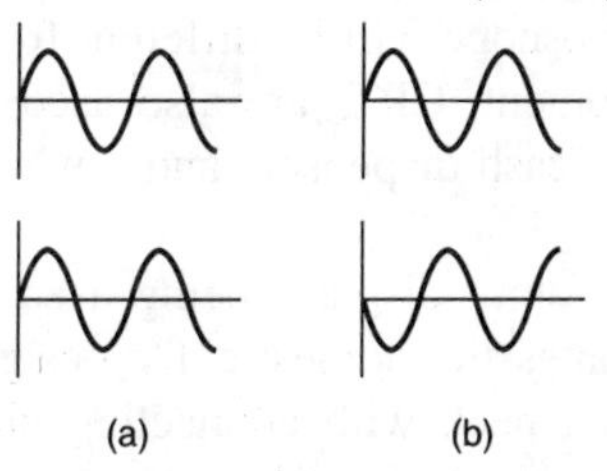

Figure 8.4 (a) Sine waves in phase. (b) Sine waves 180° out of phase

Some systems require not only a specific frequency but also a specific phase relationship. The phase of a signal is the fractional part of a period through which the signal has advanced, measured from an arbitrary origin. In the case of a simple sinusoidal quantity, the origin is usually taken as the last previous passage through zero from the negative to the positive direction. Some simple phase relationships are shown in Figure 8.4.

One cycle of a periodic waveform, a sine wave for example, is considered to travel through 360°. An ordinary single stage transistor amplifier will cause a phase shift of 180°. This is sometimes known as signal inversion (see also Chapter 10), as shown in Figure 8.5.

The transmitted chrominance (colour) information for a colour television is in quadrature, that is to say the signals are 90° out of phase with each other. In the PAL (phase alternating lines) colour television system used

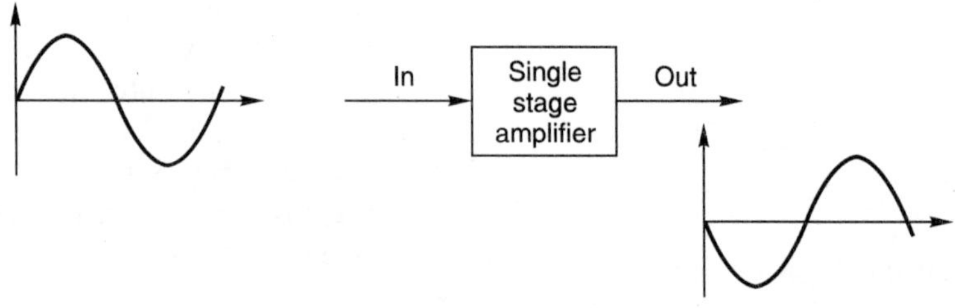

Figure 8.5 Phase inversion in a single stage amplifier

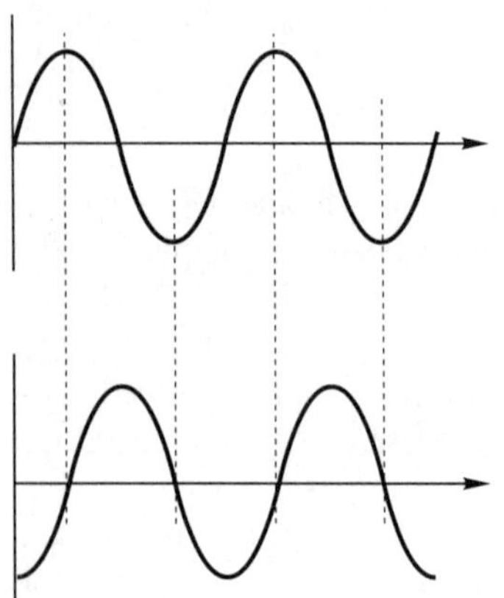

Figure 8.6 Signals in quadrature

in the UK, phase relationships are very important. The phase alternation helps to reduce colour errors, and so, not only must the signal frequency be accurate, but so too must its phase. Signals in quadrature are shown in Figure 8.6.

The detection of phase errors is accomplished by the use of a phase discriminator. From a very simplistic point of view, the phase discriminator is similar to the differential amplifier, in that the output is related to the difference between the two input signals. The differential amplifier detects voltage differences, and the phase discriminator detects differences in phase. Its output, therefore, is a function of the phase difference between its two inputs. This signal is fed to a voltage controlled oscillator (VCO), an oscillator whose output frequency depends upon a DC control voltage fed to its input. As the phase relationship varies, the output signal varies causing the VCO to increase (or decrease) in frequency until the correct phase relationship is obtained once again. The complete system is shown in Figure 8.7.

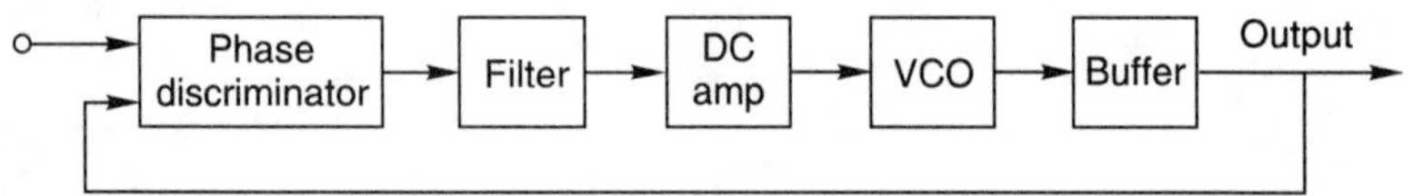

Figure 8.7 Block diagram of a phase locked loop (PLL)

The cathode ray tube (CRT)

The precise design of a cathode ray tube depends upon what it is being used for. Tubes designed for use in an oscilloscope will be different from those used in a television or in radar equipment. CRTs are also used in computer VDUs, video game displays, bank cash dispensers and a whole variety of equipment used in hospitals.

Although there are now some tiny LCD colour displays being used in portable TV receivers, the CRT still dominates the domestic TV design. Whatever the use, the CRT inevitably has a neck with connecting pins at one end containing the electron gun – a method of producing, accelerating and focusing a stream of electrons. At the other end is the screen itself whose inside is coated with a phosphor which emits light (often either green or white) when bombarded by electrons. In the case of a

CRT designed for a colour TV there will be three phosphors capable of producing red, green or blue light respectively when the electron beam strikes them.

Deflecting the cathode rays

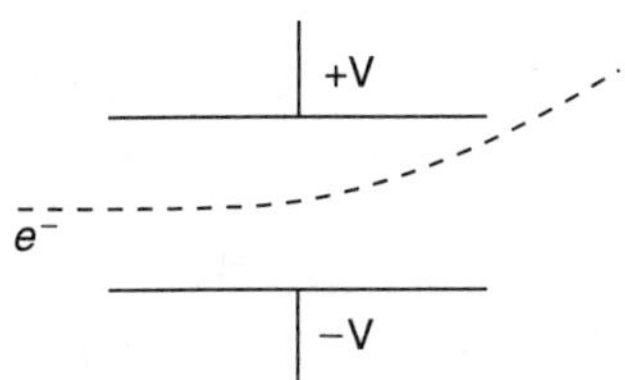

Figure 8.8 Electrostatic deflection of electrons

Electrons are tiny, negatively charged particles which surround all atoms. They are emitted from the cathode when the latter is heated. A grid then controls the number of electrons in the beam and a cylindrical anode attracts them in a certain direction and causes them to accelerate.

If the electrons are then caused to pass through metal plates across which a p.d. exists, the electrons will be deflected away from the negative plate and towards the positive plate. This is shown in Figure 8.8.

It can be shown that the electric field produces parabolic deflection. It is clear to see that two sets of plates, at right angles to each other, can produce deflection of the beam in the X and Y directions. They are therefore called X and Y plates. This is the deflection system most commonly used in oscilloscopes. Electrostatic deflection is very sensitive and can be used to display the very high frequency signals often measured by a CRO. The X and Y plates must be placed in different positions along the tube which makes it longer than it would be if this were not the case; however, this is not a severe limitation in a workshop instrument and it is useful to have the scanning mechanism inside the tube. The vertical and horizontal deflections are limited by the use of X and Y plates, but since large traces are not required, this too is of no consequence.

Electromagnetic deflection

For domestic TV applications, the screen size is always much larger and it is an advantage to have as short a tube as possible so that it is easier to build it into the popular, slimmer TV set. For these reasons, electromagnetic deflection is used for TV scanning; the scan coils can provide both horizontal and vertical deflection from the same position on the tube neck. Nor is the reduced sensitivity critical, since, while the CRO needs to scan at up to 20 MHz or more, the TV line frequency is only 15 625 kHz.

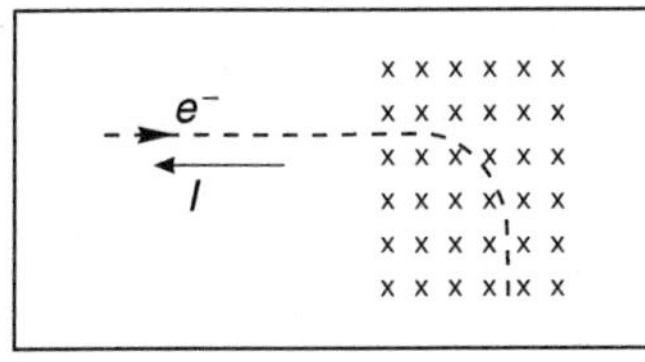

Figure 8.9 The circular deflection of electrons in a magnetic field

If a beam of electrons travels into a magnetic field, a force is exerted, the direction of which is given by Fleming's Left Hand Rule, as we saw in Chapter 4. This states that if the first and second fingers and the thumb of the left hand are held mutually at right angles, then the first finger gives the direction of the magnetic field, the second finger the direction of the current and the thumb the direction of the force. This is shown in Figure 8.9.

In the figure, the crosses indicate the magnetic field, B, going into the page (like the flight of a dart going away from you). Remembering that Fleming's Left Hand Rule requires the use of conventional current flow (opposite in direction to the actual flow of electrons) then it is easy to see that the path taken by the electron beam will be similar to that shown in Figure 8.9.

It will also be seen that vertical deflection is caused by a horizontal magnetic field and horizontal deflection by a vertical magnetic field. Unlike the parabolic deflection produced by an electric field, the magnetic field causes circular deflection.

The cathode ray oscilloscope (CRO)

The main component in any oscilloscope is the cathode ray tube (CRT) which displays the waveform being analysed. What is seen on the screen is called a 'trace' and this is produced by the bombardment of electrons on the inside of the tube screen. The screen is coated with a chemical (a phosphor) which produces green light (usually) when electrons impinge upon it. It is the kinetic energy of the electrons which is changed into light energy.

We saw way back in Chapter 1 that electrons are tiny, negatively charged particles which surround all atoms. At the opposite end of the screen, the tube contains a cathode (k) which is directly or indirectly heated causing electrons to be emitted, a process known as thermionic emission. A grid, placed close to the cathode, controls the number of electrons leaving the electron 'gun' and hence the brightness of the beam.

The beam then travels through a system of anodes which focuses the beam and accelerates it towards the face of the tube. When the beam hits the tube face, the kinetic energy of the electrons ($\frac{1}{2}m_ev^2$) is converted into light energy. A typical oscilloscope tube is shown in Figure 8.10.

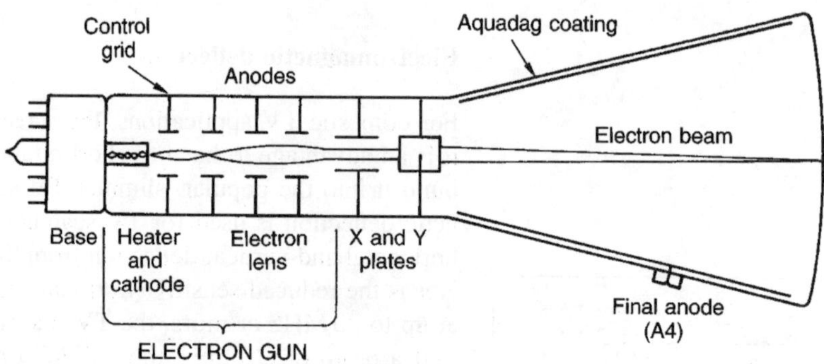

Figure 8.10 A typical oscilloscope tube

A spot on the screen is of very limited use, so it is made to sweep across the screen in order to produce a line. This is achieved by the use of a ramp or sawtooth oscillator applied to a pair of plates called X plates. The waveform is shown in Figure 8.11.

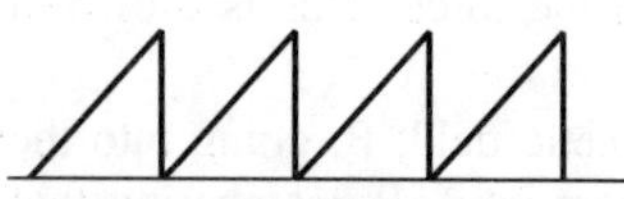

Figure 8.11 A sawtooth or ramp waveform

Since the electron beam consists of a stream of negatively charged particles, a local negative charge will repel the beam while a positive charge will attract it. The beam is therefore made to trace a line across

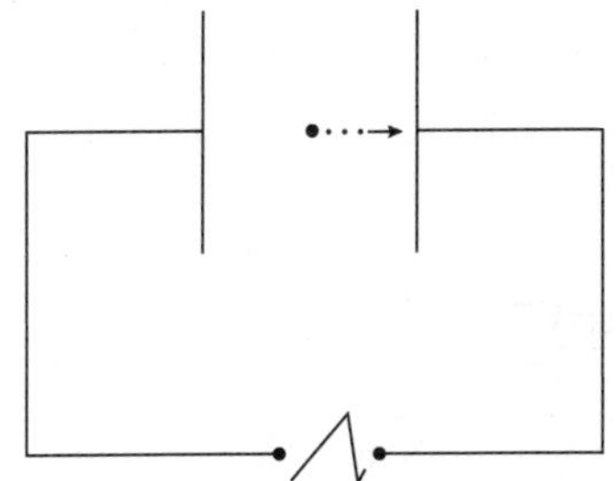

Figure 8.12 Looking at the face of the CRO tube; the spot moves from left to right and then quickly 'flies' back to repeat the action continuously

the screen by feeding a sawtooth waveform to the X plates. The oscillator providing the sawtooth waveform is called a timebase.

Looking towards the tube face as in Figure 8.12, it is possible to see how the electron beam is attracted and repelled as it moves backwards and forwards across the screen.

The timebase waveform causes the beam to move from left to right at a steady rate; this is called the sweep. The rapid return (which is made extremely small compared to the sweep) is called the flyback.

If the Y plates, which are placed at right angles to the X plates, are now fed with a suitable signal, the scanning beam will be deflected towards it, or away from it producing the waveform we wish to analyse. If a sine wave is applied to the Y plates it will appear on the screen as shown in Figure 8.13.

The flyback trace is shown emphasised in the figure, but generally, of course, it is not seen as there are circuits within the oscilloscope which suppress it. We have seen that the system of deflection using X and Y plates is called electrostatic deflection, whereas domestic televisions use magnetic deflection; there are pros and cons in each case as summarised in Table 8.1.

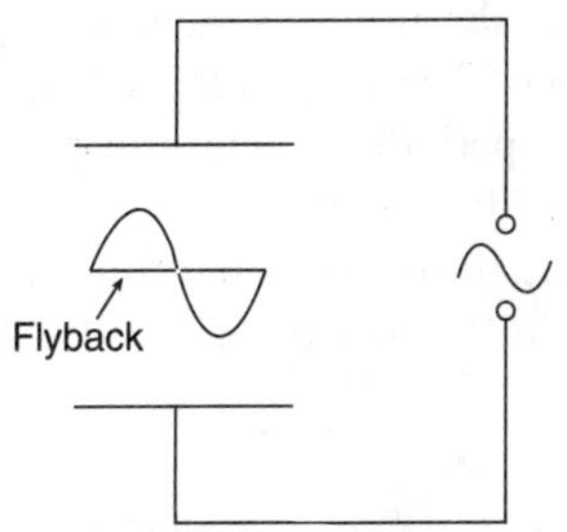

Figure 8.13 The CRO tube displays the sine wave connected to the Y plates

Table 8.1 Characteristics of electromagnetic and electrostatic deflection systems for cathode ray tubes

	TYPE OF DEFLECTION	
Characteristic	*Electromagnetic*	*Electrostatic*
length of tube	shorter	longer
frequency range	limited	very high
tube voltages	relatively high	relatively low
deflection angle	can be very large	usually quite small

In a CRO, the length of the tube is of little importance, while the frequency range must be very high. Most common oscilloscopes can handle signals up to about 20 MHz; more sophisticated machines can typically display signals up to 60 MHz and the most specialised instruments can even exceed that value. Because of the size of the tube in a CRO, accelerating voltages may be kept relatively low and for the same reason (small screen size), a large deflection angle is unnecessary.

Compare this to the requirements of a domestic TV: the tube should be kept as short as possible to provide a slimmer TV cabinet; the frequency of operation need only extend to the TV line frequency – a mere 15 625 Hz; tube voltages tend to be very high (around 25 kV) because of the large size of the tube and the deflection angle needs to be very large. Figure 8.14 shows the once common 90° deflection angle and the reduced tube length produced when the deflection angle is increased to 110°.

From these considerations, it is easy to see why electrostatic deflection is used in CRO tubes and magnetic deflection for TV tubes.

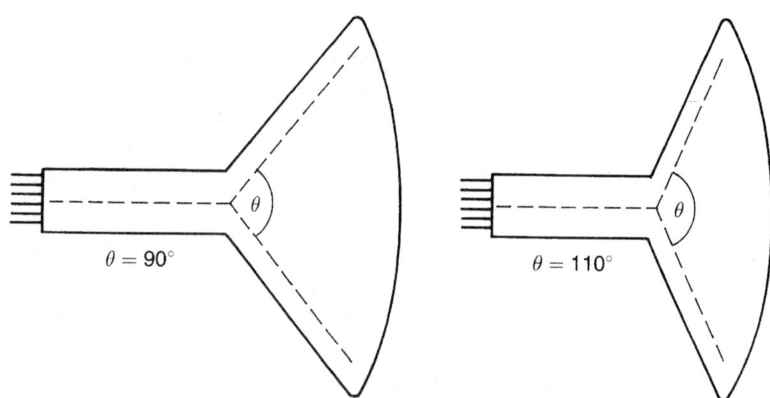

Figure 8.14 Television tubes with different deflection angles

Synchronising

The simple method of scanning described would certainly produce a 'picture' on the tube screen, but in order for the trace to be studied, it most remain stationary. This is achieved by the use of synchronising circuits often labelled 'trig' on the oscilloscope. The trig circuit ensures that the scan of an incoming signal (which is periodic) always starts in the same place and therefore remains stable on the screen.

The synchronizing circuit triggers (hence the term 'trig') the timebase oscillator which produces the horizontal scan in the first place.

Attenuation and amplification

All signals entering the oscilloscope are attenuated (reduced) and then amplified so that they can be calibrated. This gives the Y amp or AMPLITUDE control and enables incoming signals to be measured for amplitude. The timebase feeds the X amplifier which is connected to the X plates themselves; since the timebase has a variable frequency, different frequency waveforms can be displayed and analysed. Frequently there is provision for an external X timebase to be connected; this facility enables Lissajous figures to be produced.

Z modulation

As well as having external and internal X and Y inputs, some oscilloscopes have a Z connection which enables the intensity of the beam to be varied. We look at this more closely in the next section.

A complete system

A block diagram of a simple oscilloscope system is shown in Figure 8.15.

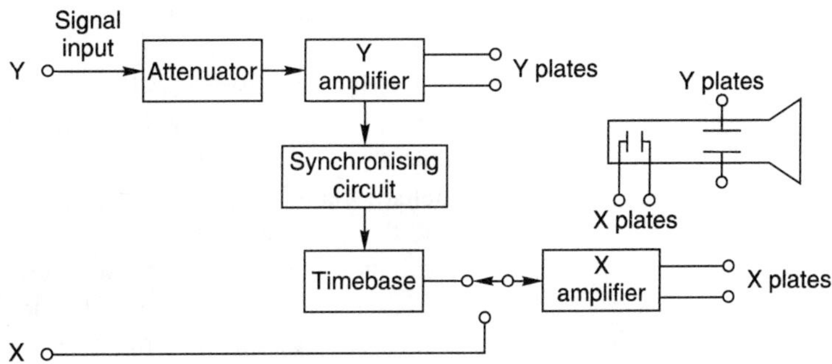

Figure 8.15 Block diagram of a simple oscilloscope system

Television systems

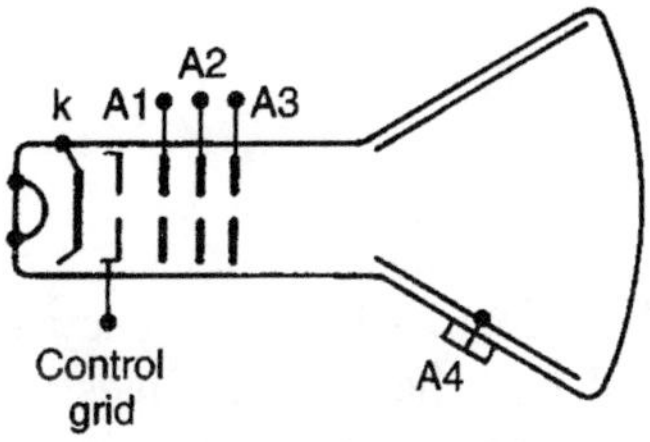

As with the oscilloscope, the heart of any TV is the cathode ray tube. Looking at monochrome (black and white) TV first, it will be seen that the neck of the tube consists of an electron 'gun' and an electron 'lens'. There are no X or Y plates since deflection is accomplished electromagnetically by the use of scan coils mounted on the tube neck. This was summarised in Table 8.1 and a typical monochrome TV tube is shown in Figure 8.16.

Figure 8.16 A typical monochrome television tube

Apart from 12″ portables, most televisions sold nowadays are in colour. When monochrome televisions were very popular the Mullard type A31-120W CRT was common. This is shown in Figure 8.17.

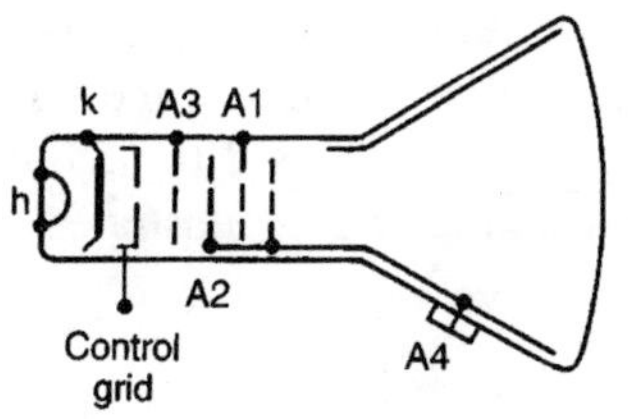

Figure 8.17 The Mullard A31-120W monochrome CRT

Typical electrode voltages are shown in Table 8.2, with corresponding voltages for a colour tube (which we look at later). Actual voltages will vary from one tube to another and will also depend upon the particular application, but the figures given represent a good guide.

The video signal is usually connected to the cathode(s) and the brightness control to the control grid. A3 is the focus anode; sometimes the brightness is controlled from the cathode.

Applying the typical voltages described to a CRT would only produce a spot on the screen, as with the CRO. The first stage in producing a picture is to scan the spot across the screen. However, unlike the CRO, a single trace is of no use, a complete picture consisting of several lines has to be built up. The complete picture consists of 25 frames per second, each frame consisting of 625 lines. The line frequency is therefore $25 \times 625 = 15\,625$ Hz.

Table 8.2 Typical electrode voltages for monochrome and colour tubes

	Heater	*Cathode*	*Control grid*	*A1*	*A3*	*A4*
Mono	6.3 V	50–80 V	−30 V to + 20 V	300 V	350 V	10 kV–17 kV
Colour	6.3 V	about 150 V	1 to 2 kV	1 to 2 kV	7 to 7.6 kV	25 kV

Interlace scanning

Pioneering television engineers found that their systems suffered from irritating picture flicker. One method used to reduce this was to employ interlace scanning. This involves the division of each frame into two 'fields' where only half the 625 lines are scanned per field. Hence, only half the lines are scanned in 1/50 of a second, first the odd numbered lines, then the even numbered ones. This system is still in use today such that each frame consists of two fields, each with a repetition rate of 50 per second, giving a frame scan of 25 Hz. Before continuing, it should be pointed out that modern TV systems do not display all 625 lines on the screen. The first few lines carry teletext information and the picture is adjusted so that those lines cannot be seen. However, for simplicity, we will ignore this and assume that a picture is still made up of 625 lines. When all 625 lines are displayed on the screen, in the absence of any video information, this is called a raster and we now turn to methods of producing the scan.

Scanning and sync circuits

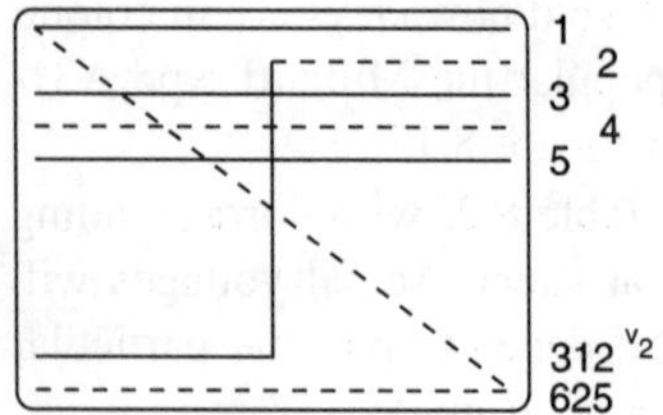

Figure 8.18 Television interlaced scanning

It has been shown that the present British UHF 625 line frequency is 15 625 Hz made up of 25 frames per second each having 625 lines (or 50 fields per second, each having $312\frac{1}{2}$ lines). The scanning waveform is a ramp (also called a sawtooth) providing a line scan and rapid flyback period equal to 1/15 625 seconds, which works out to 64 μs.

The frame scan works in a similar way except that the frame generator works at a much lower frequency, i.e. 50 Hz. After 1/50 of a second, $312\frac{1}{2}$ lines have been scanned. The frame generator has moved the trace from line 1 to line 312. In the next 1/50 of a second, lines 2 to 625 are being scanned. This is known as interlace scanning and is illustrated in Figure 8.18.

Line and frame scanning

Both line and field scan circuits have to be synchronised with the transmitter so 'sync' pulses must also be transmitted along with the picture information and we will see later how this is achieved.

The coils around the tube neck which produce the horizontal and vertical scanning by electromagnetic deflection are called 'scan coils'. Quite a lot of power is required to drive these scan coils so that as well as needing line and frame oscillators, output amplifiers are also required. The frame output is relatively low power while the line output is rather larger. Also, the EHT (extra high tension or simply 'extra high voltage') for the CRT final anode is usually obtained from the line output stage so that it has a dual function. Combining the circuitry we have described so far produces a simple block diagram for a typical monochrome TV receiver as shown in Figure 8.19.

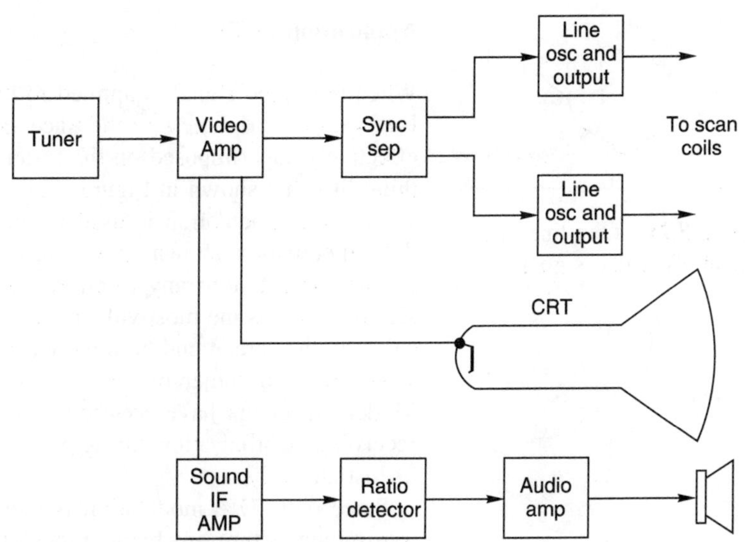

Figure 8.19 Simple block diagram for a monochrome television receiver

The block diagram

This shows the line and frame oscillators, with their power amplifiers ('output'). The line output is taken to the line scan coils and the EHT rectifier which feeds the CRT final anode. The synchronising pulses mentioned earlier derive from the video waveform and are obtained by passing the video waveform through a sync separator. The video itself is passed to the cathode of the CRT (usually) and this form of 'cathode modulation' can be demonstrated on a CRO which has a 'Z modulation' input.

Cathode or 'Z' modulation on the CRO

It will be recalled that the source of electrons in a CRT is usually from an indirectly heated cathode. By changing the potential on the cathode, more or fewer electrons will leave it, hence varying the brightness of the picture on the CRT face. The precise method used may vary from one system to another, but this model is adequate for our purposes.

Normally on a CRO the cathode is fed from a constant potential controlled by the brightness control on the CRO. However, on certain machines, the beam can be modulated in order to vary the brightness by using the Z input. If a phase shift circuit is constructed as shown in Figure 8.20 and the appropriate connections made to the X and Y inputs of the CRO in X – Y mode, an approximate circle will be produced on the screen.

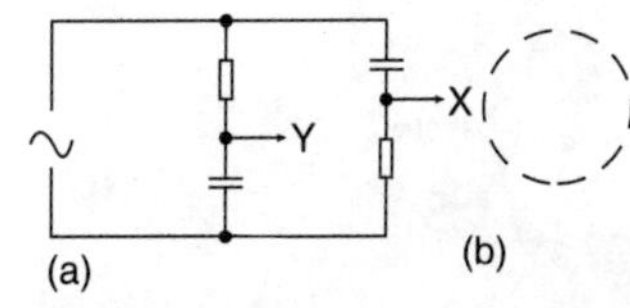

Figure 8.20 (a) A phase shift circuit. (b) A 'picture' produced on an oscilloscope

By imposing a square wave signal of an appropriate amplitude onto the Z input of the CRO, a 'picture' consisting of a broken circle will appear on the screen.

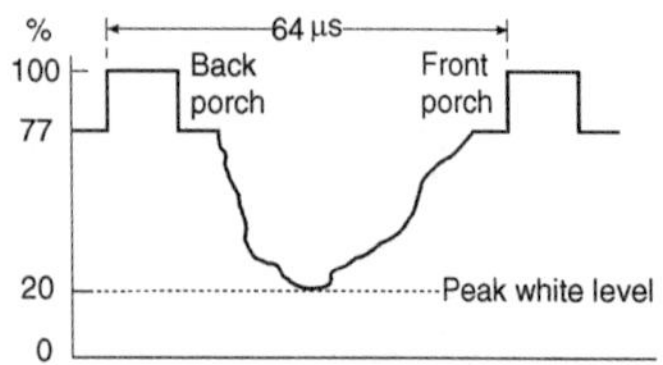

Figure 8.21 One line of a typical TV video signal

Application to TV

When the video signal is applied to the cathode of a TV tube, the result is to vary the brightness of the trace and hence produce a picture which is effectively superimposed on the raster. The video waveform looks something like that shown in Figure 8.21.

Negative modulation is used in the UK system which means that the 77% modulation shown in Figure 8.21 is in the black portion of the picture information; any interference superimposed on that part of the signal (which is the most vulnerable) would therefore show itself as black spots on the screen and be unnoticeable. Although the sync pulses themselves are also vulnerable to interference spikes, this is rarely a problem. Modern receivers have a circuit called 'flywheel sync' which maintains receiver synchronisation during periods when the sync pulses are distorted, or lost altogether.

Since only 77% modulation is required for the blackest portions of a transmitted picture, any higher modulations (where only the sync pulses lie) are called 'blacker than black'. Peak white level is set at 20% modulation.

The sync pulses occupy the region between 77% modulation and 100% modulation as shown; the leading and trailing edges of the sync pulse are called the front porch and back porch respectively.

The signal circuits

These are the vision and sound channels of a TV. In a television, the audio information is extracted from the video amplifier. The video and audio IF frequencies are separated by 6 MHz, so an amplifier – called an intercarrier amplifier – is tuned to that frequency. Since TV sound is frequency modulated, an FM demodulator is required and from then on the sound system is the same as for an ordinary FM receiver.

Summary

There are three basic requirements which must be met by a TV receiver:

1 Processing of the video signal.
2 Processing of the sound signal.
3 Production of a raster:
   (a) frame oscillator and output
   (b) line oscillator and output
   (c) EHT
   (d) synchronising pulses.

Figure 8.19 shows a full block diagram of a monochrome TV receiver. This is given for reference only and as a neat summary of what has been discussed so far. We now need to look at some basic colour television principles; it is the full colour TV block diagram with which you need to become familiar.

Colour television

One of the greatest difficulties in setting up colour television transmission in the UK was something called compatibility. This means that a colour television must be able to pick up and display a black and white picture while the many existing monochrome TVs must be able to obtain and display a black and white picture from the new colour TV signals. Before looking at this, we will take another look at some basic theory.

Sidebands

It has already been shown that when a carrier wave at a certain frequency, f_c, is amplitude modulated by a signal at another frequency, say, f_s, the resultant output contains three different signal frequencies, the carrier frequency itself, f_c, the carrier plus the modulating signal frequency, $(f_2) = f_c + f_s$, and the carrier frequency minus the signal frequency $(f_1) = f_c - f_s$. These are called the upper and lower side frequencies as shown on page 109.

This assumes that the carrier is modulated with a pure sine wave as the signal frequency. If other frequencies are also transmitted, these too will have side frequencies, and if the signal is an audio waveform, side frequencies will be continuously created at the maximum and minimum frequencies and all those in between.

When there is a band of frequencies like this the term sidebands rather than side frequencies is used. Since the range of human hearing lies between 20 Hz and 20 kHz, any audio transmission will contain sidebands with those upper and lower limits and any of those in between. The bandwidth is simply the highest frequency in the band minus the lowest frequency, for audio frequencies being 20 000 − 20 which is approximately 20 kHz and, since AM transmissions contain two sidebands, the total bandwidth will be twice the bandwidth of one sideband, in this case about 40 kHz.

UK AM broadcasts

As an example, if the carrier frequency is 1 MHz (about the middle of the MW band) and an audio signal in the audio range of 20 Hz to 20 kHz provides the modulation, then the transmitted radio waves will extend over a range from 1 MHz ± 20 kHz, that is to say 980 kHz to 1.02 MHz. So, the transmitting channel must have a bandwidth of 40 kHz in order to accommodate the LSB and USB. This is shown in Figure 8.22.

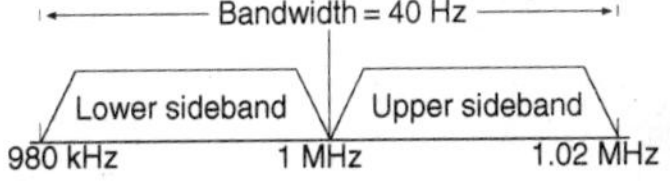

Figure 8.22 The bandwidth of a typical signal

TV reception

The medium waveband extends from about 500 kHz to 1.5 MHz so that there is only a limited amount of space available and it would not be possible to use it for TV transmissions. The UK system requires a bandwidth of about 5.5 MHz which is many times larger than the entire MW band. So, much higher frequencies (in the UHF region, in fact) are used. In addition we have already noted that AM transmission techniques produce two sidebands each carrying identical information and that many amateur radio operators eliminate one of them and use SSB – single sideband transmission. Since terrestrial TV broadcasts use AM for the video channel, why not simply omit one of the sidebands and halve the bandwidth?

Vestigial sideband (VSB) transmission

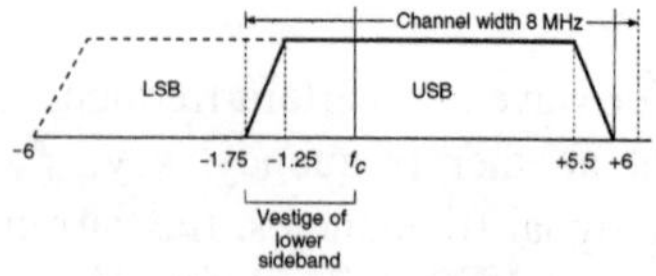

Figure 8.23 The principle of vestigial sideband transmission

The only problem is that the filtering process provides a gradual reduction in amplitude and if the whole of the lower sideband were to be eliminated, filtering would need to begin in the upper sideband and this would cause the loss of the lower video frequencies, resulting in poor picture quality. So instead, filtering begins in the lower sideband leaving part of the LSB intact. The bit left is called a vestige and so the system is called vestigial sideband transmission or VSB. This is shown in Figure 8.23.

The vision bandwidth

From Figure 8.23 it can be seen that the total bandwidth required for VSB is 8 MHz. This includes a narrow 20 kHz channel for the audio and an approximate 0.5 MHz gap before the next TV channel. Because a television picture is scanned line by line the transmitted signal consists of video information grouped around the harmonics of the line frequency of 15 625 Hz. Because of this, it is possible to transmit the colour – or chrominance – signal by using the gaps between the line frequency harmonics and, most importantly, within the existing bandwidth. This technique is known as frequency interleaving.

The standard 'black and white' TV signal is usually called the luminance signal and the chrominance signal is 4.43361875 MHz above it. This value comes from:

$$\left[\left(284 - \tfrac{1}{4}\right) \times (\text{line frequency})\right] + \left(\tfrac{1}{2}\ \text{field frequency}\right)$$

$$(283.75 \times 15625) + 25 = \underline{4.43361875 \times 10^6}$$

Because the frequency chosen is an odd multiple of half the line frequency, and it appears towards the high frequency end of the video spectrum, interference is kept to a minimum.

It was noted in Chapter 5 that a good approximation to white light may be obtained by the combination of the primary colours red, blue and green. These are the three colours used in colour TV but, in fact, only the red and blue signals are transmitted, the green signal being extracted at the receiver end by reference to the luminance and chrominance signals, in part of the circuitry known as the colour decoder.

The two chrominance signals are transmitted in quadrature (90° out of phase) which alternate in phase in order to cancel out colour errors. This system is called phase alternating lines (PAL) and is the system in use in this country.

To obtain a colour picture on the screen of a cathode ray tube requires the three primary colours. In the shadow mask tube coloured phosphors are deposited on the inside of the tube such that each will produce the appropriate coloured light when bombarded with electrons. The 'mask' referred to ensures that the electrons carrying a particular colour information, say the red, only land on the red phosphors. Some setting up of the receiver is always required and this is provided by the convergence circuits.

Modern TVs often employ alternative techniques and clever design has reduced the necessity for the complicated convergence circuits which appeared in early colour TVs. Where they exist, however, they are fed by signals from the field and line timebases. All of this information is summarised in the block diagram of Figure 8.24.

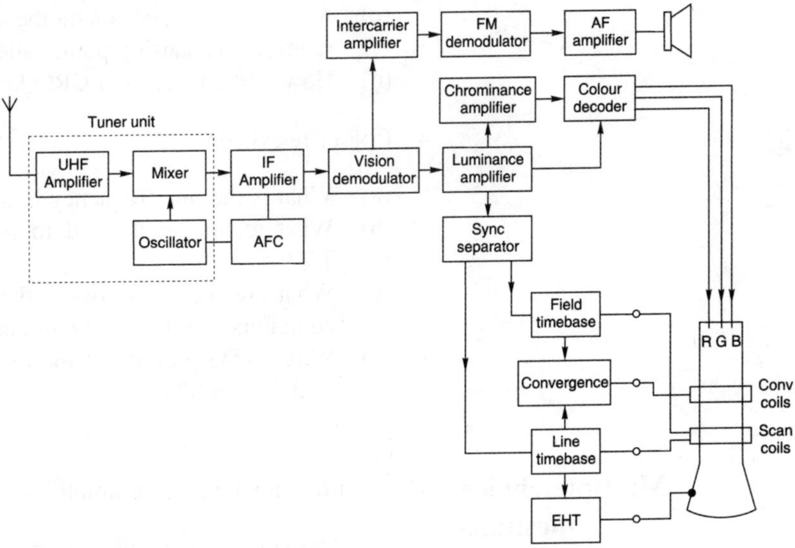

Figure 8.24 Complete block diagram of a colour television

The outline notes given in this section are necessarily limited; although colour televisions in this country are based on one basic system (PAL) there are many different versions and the technology is rapidly advancing, introducing new designs all the time. All the information required at Level 1 (and more) is given here; if you wish to know more you should seek a text which specialises in TV engineering.

Questions

1 Speed control system

(a) How is the output from the DC amplifier varied?
(b) How can the speed be automatically maintained at a preset value?
(c) What is a differential amplifier?
(d) Describe a simple method of implementing the 'set-speed' control in a simple system.

2 Phase locked loop

(a) Describe the basic function of a phase discriminator.
(b) What is a VCO and where does the control voltage come from in a PLL?
(c) What is the function of the filter in the circuit?
(d) What is the function of the buffer in the circuit?

3 The oscilloscope

(a) How is the trace produced on the screen of a CRO?
(b) What is the term used to describe the process by which electrons leave a heated cathode in a cathode ray tube?
(c) After the trace appears on the screen of a CRO, what is the rapid return to its starting point called?
(d) How is the trace on a CRO kept stationary?

4 Colour television

(a) What is the line frequency in a colour TV and how is it derived?
(b) What technique is used to reduce picture flicker in a colour TV?
(c) What are the coils around the CRT neck, which produce horizontal and vertical scanning, called?
(d) What is the purpose of the intercarrier amplifier in a colour TV sound system?

Multiple choice questions

1 A differential amplifier amplifies

A the voltage on its non-inverting input
B the voltage on its inverting input
C the sum of the two inputs
D the difference between its two inputs

2 In a speed control system the DC generator is connected to

A the output of the DC motor
B the input to the DC motor
C the output of the DC power amplifier
D the input to the DC power amplifier

3 In a phase locked loop (PLL) the block which immediately follows the VCO is

A a phase discriminator
B a filter
C a DC amplifier
D a buffer

4 The output of a phase discriminator depends upon

A the voltages applied to its inputs
B the phase relationship between its two inputs
C how well the inputs are filtered
D the bandwidth of the DC amplifier

5 The deflection system most often used in an oscilloscope is

A electrostatic
B electromagnetic
C electrokinetic
D electrolytic

6 As well as driving the Y plates of a CRT, the output from the Y amplifier in a CRO is connected to

A a synchronising circuit
B an attenuator
C a timebase
D the X amplifier

7 A CRO timebase produces a waveform which is

A a sine wave
B a square wave
C a sawtooth wave
D a series of pulses

8 A CRO trace is produced by the timebase causing the electron beam to move at a steady rate from left to right on the screen. The rapid return to its starting position is called

A the sweep
B the X–Y mode
C triggering
D the flyback

9 The length of the trace on a CRO screen is governed by

A the timebase oscillator
B the speed of the flyback

C the X amplifier
D the Y amplifier

10 The standard line frequency of a TV operating on the British TV system is

A 10 125 Hz
B 15 625 Hz
C 4 433 MHz
D 6 MHz

11 The type of modulation used for UK terrestrial broadcasts is

	Sound	Video
A	AM	AM
B	AM	FM
C	FM	AM
D	FM	FM

12 In a television built to UK standards, the block which follows the intercarrier amplifier is

A an FM demodulator
B the chrominance amplifier
C the luminance amplifier
D a sync separator

13 In a standard UK television receiver, the modulation which is required for peak white is

A 100%
B 77%
C 20%
D 0%

14 In a standard UK television receiver the chrominance subcarrier has a frequency of approximately

A 4.43 MHz
B 5.5 MHz
C 10 MHz
D 10.7 MHz

15 In a standard UK television receiver the convergence circuits are used to

A produce a linear scan
B eliminate errors caused by the scanning beams hitting the wrong colour phosphors
C eliminate chrominance signal phase errors
D produce a more concentrated electron beam and hence improve picture focus

9 Digital electronics and computers

In this chapter the function of a transistor is described. A transistor is an electronic component which has three electrodes designated collector, base and emitter. If a small current is applied to its base, a much larger current is caused to flow between its collector and emitter as shown in Figure 9.1. It is rather like the huge flow of water available in a domestic supply; there is enormous pressure forcing the water out but it can nevertheless be controlled by a small tap.

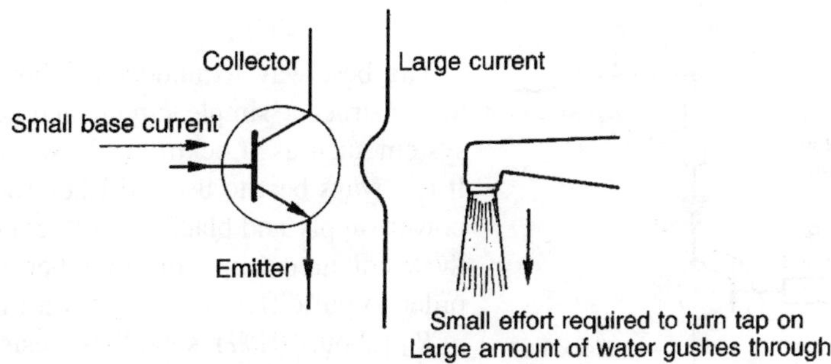

Figure 9.1 On the left, a transistor; a small current flowing into the base terminal causes a large electric current to flow from collector to emitter. On the right, a water equivalent; a small effort is required to turn the tap on, but this could cause the flow of hundreds of gallons of water per minute

In the case of the water analogy, the flow of water is continuously variable from nothing at all (its minimum value), through a small trickle, to a gushing torrent when the tap is fully open (its maximum value). This is an example of an analogue system – all the values between maximum and minimum are allowed. The transistor can also work in this fashion as we shall see in the next chapter. In an audio amplifier, collector/emitter current can be made continuously variable; the greater the current flowing into the base (the input), the greater will be the current flow between collector and emitter (producing the output).

The transistor as a switch

In digital electronics only two states are allowed – fully on (a logic 1) or fully off (a logic 0). The transistor then operates as a rapid and reliable electronic switch. Figure 9.2 shows two simple transistor switches.

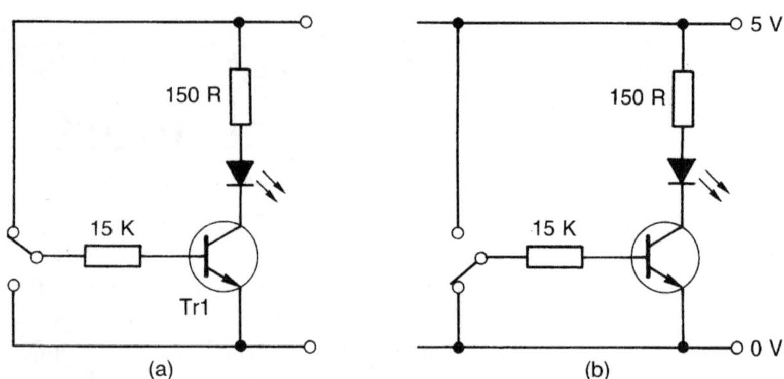

Figure 9.2 Two simple transistor switches. The LEDs show the logic states; when the transistor is 'ON', the LED glows as current flows through the collector/emitter circuit; when the transistor is 'OFF' the LED is off. In circuit (a), Tr1 is 'ON' and in circuit (b) the transistor is 'OFF'

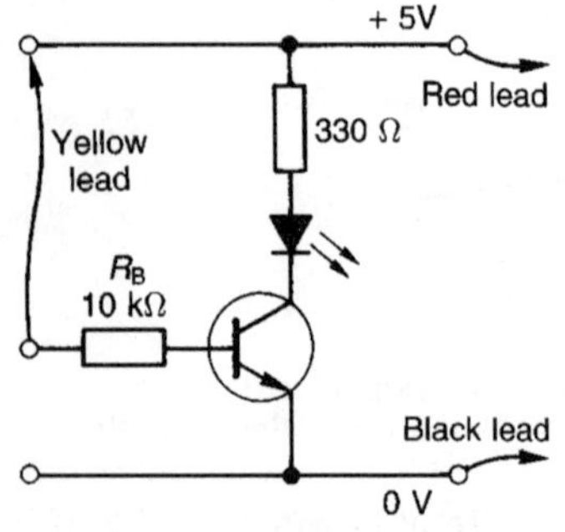

Figure 9.3 The transistor acting as a switch

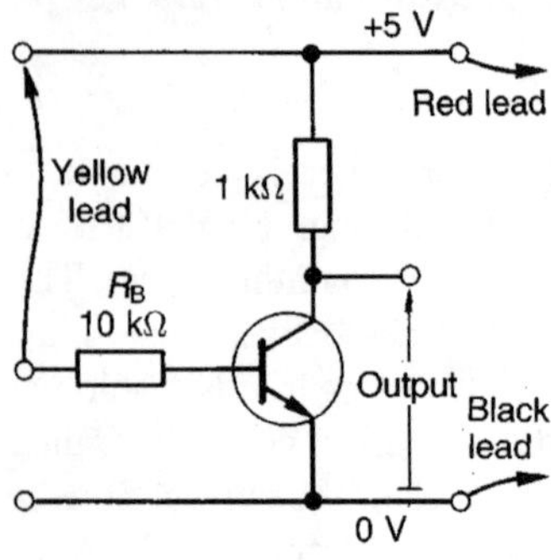

Figure 9.4 The transistor as a simple logic gate

The best way to understand how the transistor works as a switch is to construct a simple circuit and test it. This can best be done with a system such as 'Locktronics'. Set up the circuit as shown in Figure 9.3. It is always best to use red leads for the connection to the positive of the power supply and black leads to connect to the 0 V line and to avoid using these coloured wires for any other purpose. A yellow 'fly lead' is used in order to put different voltages on the input.

R_B (about 10 kΩ) is the 'base resistor'. Touch the fly lead on the positive of the supply. The LED should glow indicating that the transistor is turned ON. Connecting the base (via the 10 kΩ resistor, of course) to 0 V (or simply removing the base connection in this case) turns the transistor switch OFF.

Try larger values of base resistor and note that the circuit still functions with much higher values indicating how little current is needed in the base/emitter circuit. You may recall from Chapter 4 the touch sensitive switch in the section dealing with relays. The tiny amount of current flowing through a finger is enough to turn the transistor on.

The simple transistor switch circuit may be regarded as a basic LOGIC GATE. This is an electronic circuit whose output depends entirely upon its input. In this case, a 'HIGH' input (connecting the base, via R_B, to +5 V) produces a 'LOW' output – measure the transistor collector voltage to see. The behaviour of a logic gate may be recorded in a TRUTH TABLE, but first, replace the LED and 330 Ω resistor with a single 1 kΩ resistor, as shown in Figure 9.4, then make the appropriate tests in order to complete the truth tables shown in Figure 9.5.

A simple version of what is called a NOT gate has been produced. You should find that when the input is 'low' (0 V or a logic 0) the output is 'high' (5 V or a logic 1). The output is NOT the input. This circuit is also called an inverter.

Input	Output
0 V	
5 V	

(a)

Input	Output
LOW	
HIGH	

(b)

Input	Output
0	
1	

(c)

Figure 9.5 Truth tables for a simple logic gate. (a) gives the voltage levels you should measure on the collector, (b) shows how these levels are simply described as 'low' and 'high', and (c) gives the actual logic levels

Try having two inputs to the gate (two separate 10 kΩ resistors feeding the transistor base connection) as shown in Figure 9.6 then complete the truth table (Figure 9.7) by experiment.

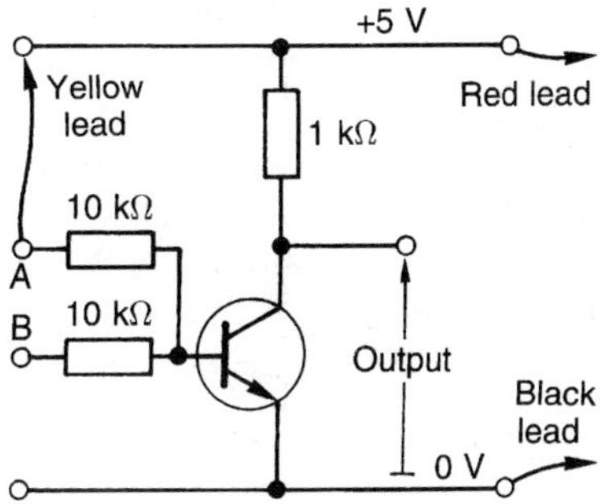

Figure 9.6 A more complex gate now with two inputs

Input		Output
A	B	
0	0	
0	1	
1	0	
1	1	

Figure 9.7 Truth table for two-input logic gate

Now connect the circuits of Figures 9.4 and 9.6 together, such that the output of the Figure 9.6 circuit becomes the input of the Figure 9.4 circuit. By doing this, we are following the NOR gate with an INVERTER and producing an OR gate. The BS symbols for this arrangement are shown in Figure 9.8.

Figure 9.8 BS symbols for a NOR gate followed by an inverter

Now draw up a truth table as shown in Figure 9.7 for the OR gate. From this you should be able to state that the output is high when either A *or* B is high. The British Standard symbol for the OR gate is the same as the NOR without the inverting 'pip' on the output.

The next step is to set up the circuit shown in Figure 9.9. This arrangement shows two transistors, one on top of the other, an arrangement known as 'cascode'. Again, there are two inputs and it is now a simple matter to draw up a truth table for this gate (using, again, the blank shown in Figure 9.7 for a two-input logic gate).

From the truth table you should note that the only way to obtain a logic zero at the output is to have both inputs set at a logic 1. This is called a NAND gate. The final experiment in this series is to follow the Figure 9.9 circuit with an inverter and then draw up the truth table by applying the appropriate inputs in turn and noting the outputs. This arrangement is shown symbolically in Figure 9.10.

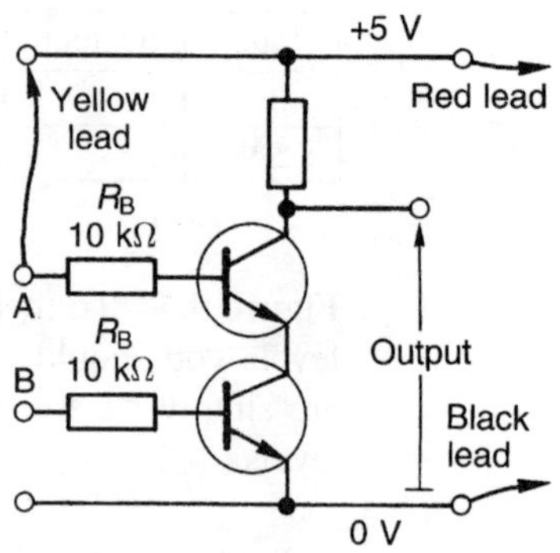

Figure 9.9 Two transistors forming a simple logic gate

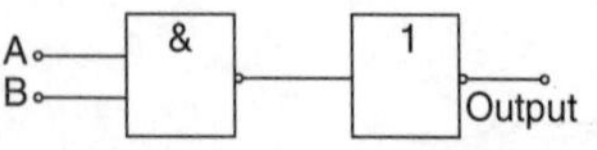

Figure 9.10 BS symbols for a NAND gate followed by an inverter

Note that these circuits are very basic and for demonstration only; manufactured gates (like the 7400 IC series) are more sophisticated, such that the devices present low impedances at their outputs for both logic levels. The American symbols for logic gates and their truth tables are shown in Figure 9.11.

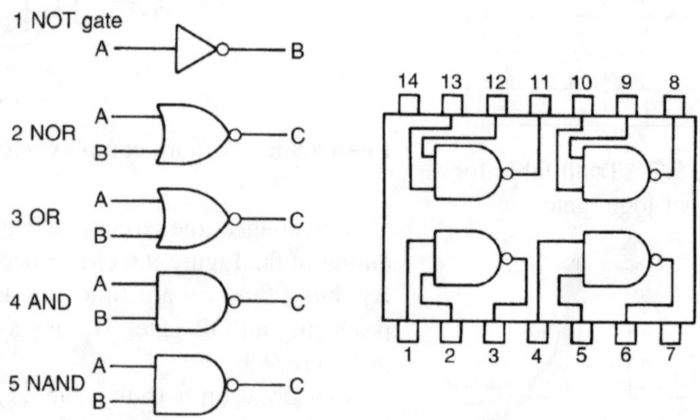

Figure 9.11 American logic symbols and (bottom) layout of 7400 quad NAND chip

Combinational logic

In order to produce some of the gates described we have followed one basic gate with another. For example, a NOR gate was followed by an inverter to produce the OR gate, and a NAND gate was followed by an inverter in order to produce the AND gate. This is a simple form of combinational logic. More complicated logic systems can be created to solve more complex problems. Figure 9.12 shows a simple, combinational logic circuit.

In this case, you have to work out what the outputs at X, Y and Z are for given inputs at A, B and C. For example, suppose that A = 1, B = 1

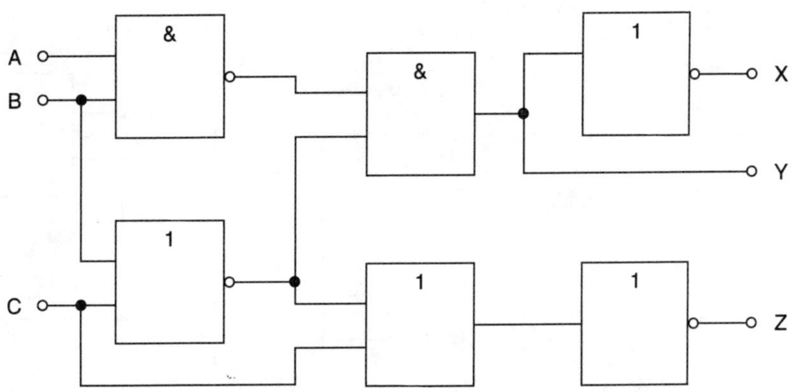

Figure 9.12 A typical, combinational logic circuit with three inputs

and C = 0, what are the outputs at X, Y and Z? One way of doing this is to examine the outputs at each gate in turn. Since the top left gate is a NAND gate, the only way to get a zero is if both inputs are at logic 1. Since they are, the output is a logic 0 which is then passed on to the next gate which is an AND. Try to work through the circuit yourself in the same way, before consulting the solution which appears in Figure 9.13.

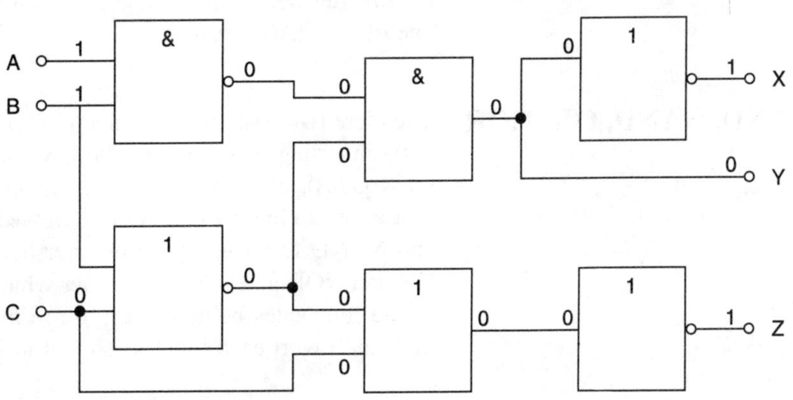

Figure 9.13 Solution to the combinational logic circuit problem given in Figure 9.12. The output is 1, 0, 1

Another simple combinational logic circuit appears in Figure 9.14. Work out the outputs at X, Y and Z, given that the inputs at A, B and C are 1, 1 and 0, respectively. The answer to this problem appears at the end of this chapter.

Inverters

If the inputs to either NAND or NOR gates are connected together, they become simple inverters. This can be very useful when designing logic circuits. We will see later that many 7400 series chips (and the CMOS

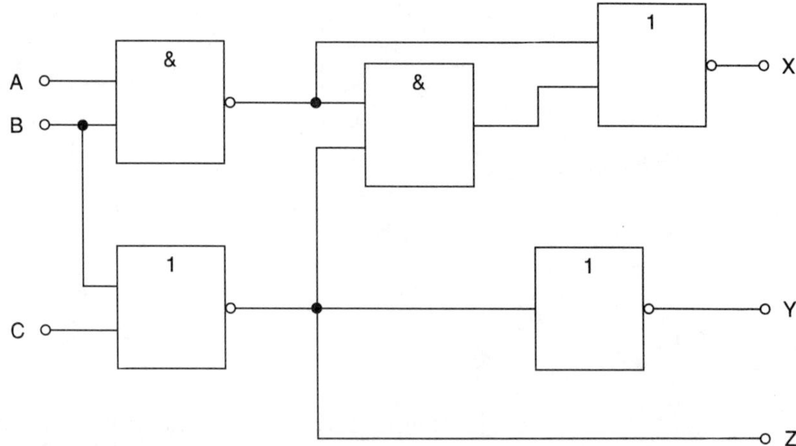

Figure 9.14 Another combinational logic circuit

equivalents) each contain four gates. The 7400 chip itself, for example, is called a 'quad NAND gate' because it contains four NAND gates. The 7408 contains four AND gates. If a circuit requires three NAND gates and an inverter, the circuit can be implemented using a single 7400 chip, the inverter being implemented by connecting together the two inputs of one of the NAND gates.

AND, NAND, OR, NOR

There are two basic ways of changing one logic gate into another, the first is by inverting its output (as we have seen in the case of the NAND and NOR gates), and the second is by inverting its inputs. If the first method is used it is obvious that AND followed by an inverter becomes NAND, and NAND followed by an inverter becomes AND. The same applies to OR and NOR gates. Now consider what happens when the *inputs* to each of the four gates being discussed are inverted. We'll start with the AND gate and invert each input as shown in Figure 9.15.

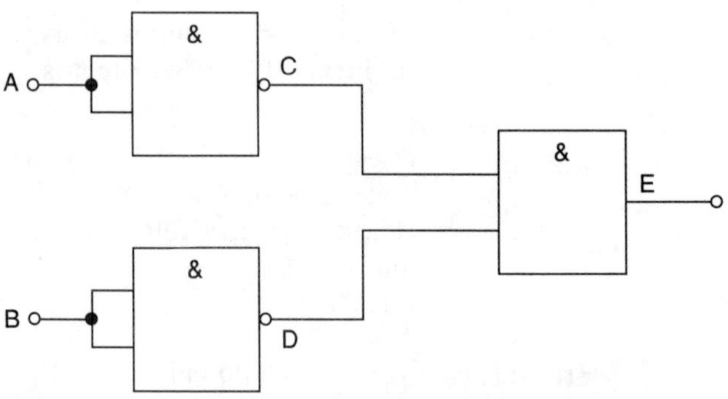

Figure 9.15 An AND gate with its inputs inverted

Examination of this circuit shows that for the four possible pairs of inputs, the outputs are 1, 0, 0, 0, which is the truth table for a NOR gate. It is helpful if a truth table with two extra columns is used, as shown in Figure 9.16.

Now draw out the NAND, OR and NOR gates with inverted inputs and work out their truth tables. Do this before consulting the solutions shown in Table 9.1.

Inputs				Output
A	B	C	D	E
0	0			
0	1			
1	0			
1	1			

Figure 9.16 An extended truth table

Table 9.1 The result of inverting the inputs to certain gates

Gate with inverted inputs	*What it becomes*
AND	NOR
NAND	OR
OR	NAND
NOR	AND

Note that in Table 9.1 AND and NAND gates are grouped together as are the OR and NOR gates. A gate from one group with inverted inputs produces a gate from the other group; noting that in each case, it's the 'complementary' gate that is produced – it's a simple matter to remember what each logic circuit will produce.

Now consider what would happen if two, simple transistor switches are connected 'back to back' and capacitor-coupled to each other as shown in Figure 9.17. So that it is possible to monitor what's happening, LED indicators are included.

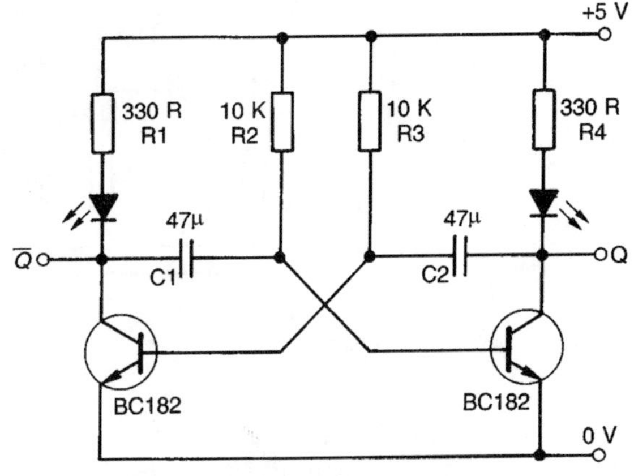

Figure 9.17 An astable multivibrator which flashes LEDs

The astable

Try setting up the circuit on a Locktronics board or similar. If the circuit is connected correctly, you should see that the two LEDs flash on and off alternately. This circuit is called an astable multivibrator. Astable means that the circuit has no stable state and continues to oscillate as long as power is applied. The word 'multivibrator' means multiple vibrations. This is because the theoretical output from the astable is a square wave, and theory tells us that square waves consist of an infinite number of harmonically related sine waves – the multiple vibrations.

If C1 and C2 are replaced by 0.1 μF capacitors, the circuit oscillates so rapidly that the LEDs seem to glow constantly. The shape of the waveform (an approximate square wave) can be seen by connecting an oscilloscope to the collector of one of the transistors.

Try removing both LEDs, replacing the left hand one with a 1 kΩ resistor and the right hand one with a small loudspeaker in series with a 1 kΩ resistor. When power is applied, the loudspeaker should emit an audible tone. The pitch of the tone can be varied by putting in different values of capacitor, and/or different values of R2 and R3. You are encouraged to experiment with this.

The bistable

The astable is a square wave oscillator with no stable state; as long as power is applied, the LEDs continue flashing, or the loudspeaker emits an audible tone. The bistable has two stable states and it requires a pulse to 'flip' it from one state to another. For this reason, the circuit is often called a 'flip-flop'. Connect up the circuit shown in Figure 9.18 and test it.

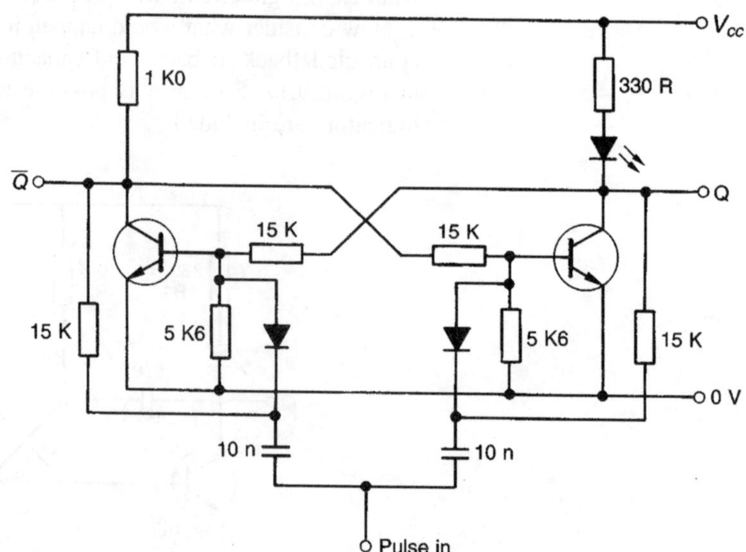

Figure 9.18 A bistable or 'flip-flop' circuit

If the circuit is connected correctly, the LED will either be on or off. The application of a suitable pulse should flip the circuit from one state to another, so if the LED were on when power was applied, the pulse should turn it off; if it were off, then the pulse should turn it on. Simply connecting and then quickly removing the pulse input from the +5 V power supply should provide a suitable pulse. If the pulse input is then connected to one of the transistor collectors of the astable, the bistable will be continuously flipped from one stable state to the other.

A 'clock' oscillator

When an astable is used to supply a pulse in this way it may be described as a 'clock' oscillator and the bistable is therefore being 'clocked' by the astable. The output from the astable rises (to about 5 V if that's the power supply) when the LED is off, and falls almost to 0 V when the LED is on. This explains why the output is an approximate square wave. The bistable is 'clocked' during the period when the astable collector voltage falls from 5 V towards zero. (This is called 'negative-edge' triggering.) At this point, the bistable takes up one of its stable states. It then does nothing while the other LED is flashing on and off and changes state again only when the astable collector it is connected to, makes its 5 V to 0 V transition. For this reason, the flash rate of the bistable's LED will be half that of the astable which is clocking it.

The bistable as a 'divider'

The bistable is therefore acting like a 'divide by two' circuit. If it's being clocked at 500 Hz, then its output will be 250 Hz. Circuits like this can therefore be used as 'dividers' in order to reduce the output frequency of an oscillator. We'll see this in action when discussing a digital clock (in this case a clock which is used to tell the time).

The bistable as a memory device

Since the bistable changes state only when a pulse is applied, that state will be retained indefinitely (in the absence of any further pulses), as long as power is applied to the circuit. In this case, since we can regard the higher voltage as a logic 1, and the lower voltage as a logic zero, a bistable can 'remember' either logic value. Once set, the bistable circuit remembers what it is storing as long as power is connected. For this reason, the bistable can act as a memory device for a computer. Yes, indeed, the whole circuit can only store one number (0 or 1), so it seems like a lot of effort to use two transistors and all the resistors, capacitors and diodes and so on that are required for the purpose. Maybe; in fact, with modern integrated circuit technology, two such circuits can be mounted inside a tiny, 16-pin integrated circuit package – the 7476, for example.

Even more impressive is the fact that computer memory chips can contain thousands, or even millions, of bistables on a tiny chip about 1 cm by 3 cm. We will see this in action when we discuss a simple computer system in a few moments.

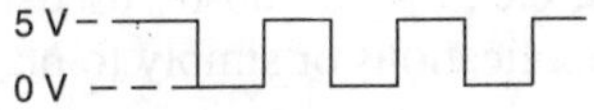

5 V
0 V

Figure 9.19 A 'square' wave. Assuming TTL logic levels, 5 V = 'ON' (logic 1) and 0V (zero volts) = 'OFF' (logic 0)

The bistable as a binary counter

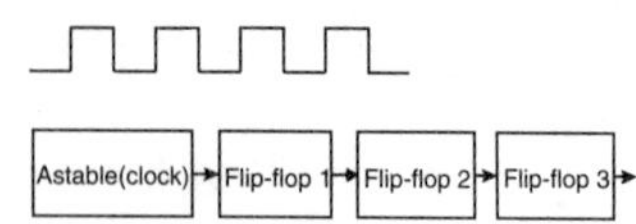

Figure 9.20 Linked bistable operation. On the first pulse from the clock, flip-flop 1 changes state. On the next pulse, flip-flop 1 returns to its original condition and emits a pulse to flip-flop 2 allowing it to change state. The effect is carried through to any following flip-flops. Placed in reverse order (FF1 on the far right), such a system 'counts' up in binary

The bistable has been described as being 'clocked' by an astable when the latter's output produces a 5 V to 0 V transition – the negative-edge triggering described before. Higher clocking frequencies can most easily be represented by a square wave as shown in Figure 9.19.

As the bistable LED turns on and off, its collector is also making similar transitions, and these pulses can be used to clock other bistables. In addition to making further divisions by two, such a system counts up in binary. This linked bistable operation is shown in Figure 9.20.

If the linked bistables are placed so that the astable clock is on the far right and each successive bistable is placed towards the left, then it is easy to see how the binary count takes place – this is shown later in this chapter in Figure 9.24. On the first clock pulse, FF1 (flip-flop 1 or bistable 1) becomes a logic 1. On the second pulse, FF1 goes to a zero and FF2 becomes a 1, this produces 10B (1, 0 in binary, 2 in decimal). On the third pulse, FF1 comes back on again (FF2 is unaffected) so the system now displays 1,1, in other words 11B or 3 in decimal. The fourth pulse turns FF1 off, FF2 off and FF3 on, the display becoming 100B, 4 in decimal. The process continues and the system counts up in binary as long as power is applied.

A computer clock

The astable clock just described is quite a primitive thing although it can quite adequately provide the pulses for a binary counter as we have seen. All computer systems have clocks to enable them to step through instructions and execute a program. For speed, the clock oscillator can operate at a rate of several million times a second, and some modern processor systems may exceed 100 MHz.

A computer system

A microelectronic system is comprised of a microprocessor (MPU – microprocessor unit – or CPU – central processor unit), suitable support chips (memory and interfacing electronics, for example), which can control an operation such as traffic lights, a washing machine, a manufacturing process or a motor car; something which has only one use or purpose. A microcomputer system is also controlled by an MPU, but in addition has a VDU (video display unit), keyboard, floppy disk drives, a hard disk, in modern systems a CD-ROM drive, loudspeaker outputs and (generally) a printer. In such a system, the user can run different programs, such as word processing, databases and spreadsheets, etc., and it can be used in order to develop and run programs for other applications or simply to play games.

The microprocessor lies at the heart of any computer. It contains a huge complex of combinational and sequential logic gates, arranged in such a way that they can carry out the control, data, address and memory functions required by a digital computer. The most common types of memory chips called RAMs (random access memory), work by the setting and resetting of tiny flip-flop or bistable circuits. These chips hold text or programs as you type, or load them into the computer; all this information is lost if power to the computer is interrupted.

A user would normally 'save' the contents of RAM (usually the program) on floppy or hard disk before turning the power off; such systems are called memory backing stores. It is possible to write a program and have it stored permanently in a ROM (read only memory) chip, but the process is quite complex and requires sophisticated equipment. ROM memory chips are usually programmed by a manufacturer; these chips contain the computer operating system and standard routines that may be used frequently. This information is stored permanently in the ROM chip and is not lost when power to the machine is switched off. It is termed a 'non-volatile' memory.

The term 'random access memory' can be applied to both RAM and ROM. Put simply, this means that any data byte stored in the memory chip can be obtained as easily as the next; it doesn't matter where in the chip it is. In the term 'ROM', 'read' simply means the accessing of information from the chip; the opposite term, 'write', means the process of putting data into the memory chip. Hence, the differences (and similarities) between RAM and ROM chips may be summarised in Table 9.2. Some other memory chip details are included for reference.

Table 9.2 ROM and RAM chips compared

Chip name	*Random Access*	*Read/write*	*On power off*
RAM	Yes	Both	Loses memory
ROM	Yes	Read only	Retains memory
PROM	Yes	Read only	Retains memory
EPROM	Yes	Both	Retains memory
EAROM	Yes	Read, but erasable	Retains memory
EEPROM	Yes	Read, but erasable	Retains memory

Interfacing

A computer is of little use if it cannot communicate with the outside world. To be useful, the system must be able to take in data via a keyboard, or (when running certain games) via a 'joystick' or similar control. The system must also be able to produce an output, typically in the form of text or pictures on a VDU, often in the form of sound (music, bleeps, sound effects and so on) and usually to produce a hard copy using a printer.

Interfaces are used in order to make a computer output compatible with the device it wants to drive; looking at it the other way, an interface is necessary to convert the output from a device into something the computer can use. In each case, the process is sometimes referred to as 'signal conditioning'.

Signals

The word 'signal' in this context means an electrical or electronic value, or variation in value, which carries information. Signals have no power, they only carry information. The television signals appearing at the end of the coaxial cable used to join the aerial on the roof to the television inside the house, are of the order of microvolts and they cannot 'drive' anything either. But all the complex information about the sound, vision and colour, together with synchronising pulses and teletext, are all carried in the signal. It only has to be processed and amplified by the television receiver (quite a complex task). You cannot simply hang a loudspeaker on the end of the aerial cable and expect to hear a sound – the signal needs to be interfaced.

Incidentally, you may hear TV engineers talking about 'signal circuits'. These are the earlier circuits in a TV consisting of the tuner, detectors and IF strip. Because this is where the signal is being processed, no great power is involved; the circuits run cool and tend to be extremely reliable. It is the output circuits which consume the power, those that produce the lines on the screen, the video and the sound.

An attempt has been made to make this section quite comprehensive since computing equipment seems to enter everyone's lives these days. There continues to be an army of technicians and engineers working on future developments and design and the repair and servicing of the plethora of microelectronic and computing systems in present day use. For our purposes, a working knowledge of a very simple block diagram is all that's required. This is shown in Figure 9.21.

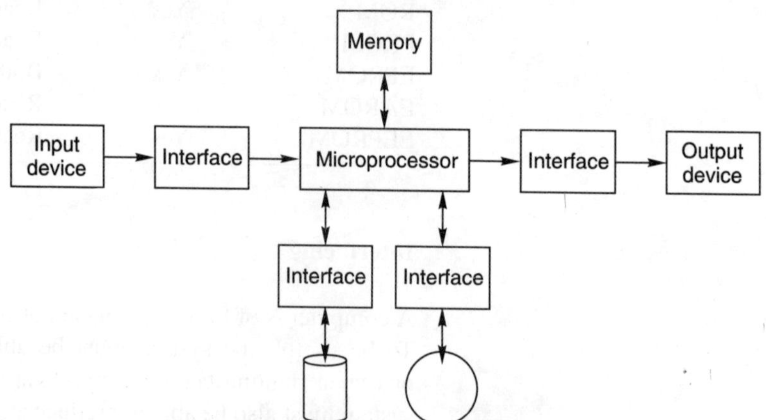

Figure 9.21 Block diagram of a computer system

If you are interested in pursuing a career in computers and/or micro-electronics you will find the information in this section of great interest and may wish to use it as a starting point for further investigations, beyond the requirements of the course at this level.

The digital clock

We have already seen that an astable multivibrator will produce a square wave pulse at a frequency determined by the values of *R* and *C* in the circuit. The circuit is called an astable because it is free running and has no stable state.

The bistable, or flip-flop, is a similar circuit but in this case has two stable states. In order to change state, a pulse is required. The bistable changes state once for each pulse applied and therefore requires two of them in order to return to its original state. If a slow-running system consisting of astable and bistable has LEDs to indicate the state of each, it will be seen that the bistable effectively divides by two. As many bistables as are required may be connected in series so that each one divides by two. A very fast-running astable can be divided hundreds of times to produce one pulse per second, one per minute, or even one per hour. This is the basis of a (real time) digital clock.

Accuracy

For driving a real-time clock, the oscillator must be very accurate and remain accurate during operation. Crystal-controlled oscillators at high frequency are generally used. The crystal ensures great accuracy and the higher frequency means that any variations become insignificant. If the master oscillator runs at about 10 MHz, the variation of a few cycles becomes unimportant, especially when the frequency is divided down to only one pulse per second.

The crystal

Certain crystalline substances (quartz in particular) exhibit what is known as the 'piezo-electric effect', i.e. a mechanical strain applied to a suitably cut piece of the substance causes a small emf to be developed between its opposite faces. Similarly, a small voltage applied between two faces of the crystal will cause mechanical deformation.

The naturally occurring crystal of quartz is hexagonal, about 7 cm in diameter with pointed ends making the crystal up to 15 cm long. The crystals may be cut in various ways so that they have precisely defined resonant frequencies. This means that when connected into an electronic circuit they vibrate at their natural frequency, which then becomes the frequency of the oscillator. The lowest practical frequency is about 20 kHz, but it's common to have crystals which resonate at 10 MHz. Because of this, the frequency has to be divided down for use in a real-time clock.

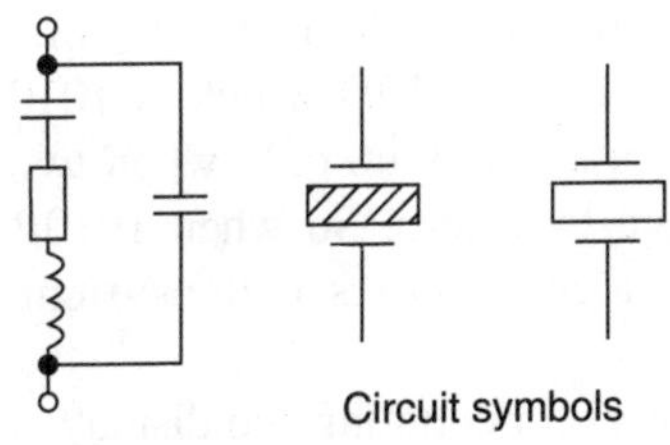

Figure 9.22 Equivalent circuit of a piezo crystal and its circuit symbols

Electrodes are formed on the crystal by depositing gold or silver on opposite faces. The connecting wires to these electrodes also help to support the crystal in its holder. The equivalent circuit of a piezo crystal and its circuit symbols are shown in Figure 9.22.

Dividing ICs are available which divide by ten or divide by six. The 7490 IC is a divide by ten, the 7492 divides by six. If the master oscillator runs at 10 MHz, the following frequencies are obtained after one and more stages, each dividing by ten:

10 MHz : 1 MHz, 100 kHz, 10 kHz, 1 kHz, 100 Hz, 10 Hz, 1 Hz

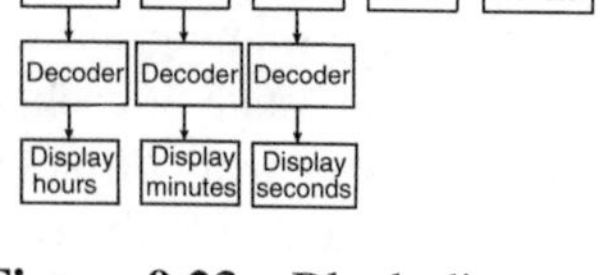

Figure 9.23 Block diagram of a digital clock

It is common to take the 1 Hz signal and pulse the full stop or colon which normally divides the hours and minutes on a digital clock. City and Guilds, however, prefer the block diagram of a digital clock system which also displays seconds. This is shown in Figure 9.23.

The frequency of a crystal can be changed or 'pulled' slightly by a few kHz by the variation of capacitance in parallel with the crystal; a common value would be about 50 pF.

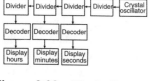

Binary coded decimal (BCD) numbers

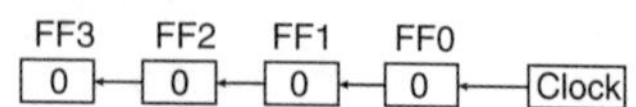

Figure 9.24 A row of four bistables being clocked. A set of four like this is capable of displaying any number between 0 and 1111B (15D)

Binary counting using linked bistables driven by a clock has already been discussed. If four bistables are connected together and clocked, they can count up to 15. This state is reached when each bistable is holding a logic 1, the four of them show 1111B which is 15 in decimal. There are many uses for such a counter, but the digital clock isn't one of them.

Figure 9.24 shows four bistables being clocked from the right hand side. Between the four of them they can display any combination of binary numbers from 0 (all zeros) to 15 (when they display all ones).

Assuming a digital clock which displays hours and minutes (e.g. 17:00 meaning 5 o'clock), each 7-segment display shows one of the numbers as in Figure 17.1 (see page 211).

Since the display is limited to one digit, it is obvious that it can only display the numbers 0 to 9. This means that the numbers 10, 11, 12, 13, 14, and 15 are 'not allowed'.

If a system is produced which counts up to 9 in binary and then returns to zero, it would be counting in BCD – binary coded decimal. This is achieved by attaching a NAND gate to flip-flop 1 and flip-flop 3 in Figure 9.24. Now, when the system counts up from 0 to 9, there is never a display where both FF1 and FF3 are both at logic 1. The first time this happens is when the counter tries to reach 10D which is 1010 in binary. The NAND gate normally has a logic 1 output, only when two ones appear on its input, does the output go to a zero. So when 1010B (10D) is reached, the NAND gate produces a zero and resets the system. This is shown in Figure 9.25.

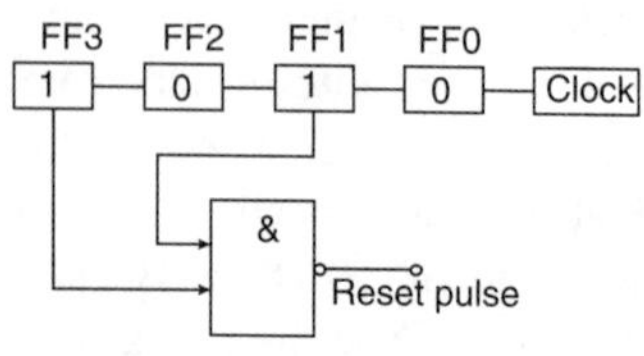

Figure 9.25 Binary counter modified to produce BCD numbers. When the binary count reaches 1010B (10D), the NAND gate produces a zero which is used to reset the system to all zeros

A digital clock, therefore, uses BCD numbers for counting and display; a 'decoder' is then used to convert the BCD number (0 to 1001) into decimal (0 to 9). This signal is then applied to a 7-segment display 'driver' and hence to the display itself. The driver is like an 'interface', it can accept the low power signal from the decoder and give it the necessary power to drive the display.

Combinational logic solution

The solution to the combinational logic circuit of Figure 9.14 is given in Figure 9.26.

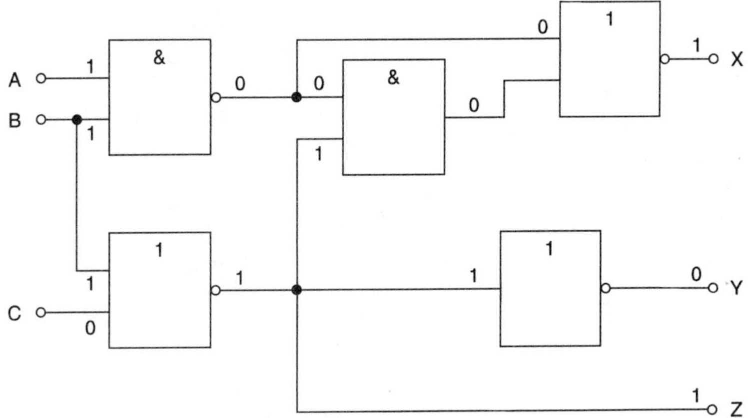

Figure 9.26 Solution to combinational logic circuit of Figure 9.14

Multiple choice questions

1 A transistor has three electrodes designated

A collector, base and current
B voltage, current and base
C base, collector and emitter
D base, emitter and voltage

2 The term 'flip-flop' is another name for

A an oscillator
B an astable
C a monostable
D a bistable

3 When using a transistor as a switch, a typical value for the base resistor is

A $1\,\Omega$
B $100\,\Omega$
C $10\,k\Omega$
D $100\,k\Omega$

4 A single transistor amplifier acting as a switch inverts any input signal, this may be regarded as a simple

A NOT gate
B NOR gate
C OR gate
D NAND gate

A	B	Out
0	0	0
0	1	1
1	0	1
1	1	1

Figure 9.27

5 The truth table shown in Figure 9.27 is for

A a NOT gate
B a NOR gate
C an OR gate
D a NAND gate

6 If the two inputs to a NAND gate are inverted, the arrangement produces

A an AND gate
B an OR gate
C a NOR gate
D a NOT gate

7 An astable 'LED flasher' has two 47 μF cross-coupling capacitors and the flash rate is about once a second. If the 47 μF capacitors were replaced with 0.1 μF capacitors the circuit would

A stop dead
B not work at all
C increase the flash rate
D decrease the flash rate

8 A simple binary counter made up from four bistables displays the binary number 0101B. After four more pulses the counter would display

A 0111
B 1000
C 1001
D 1010

9 The two inputs of a NAND gate are connected together. This arrangement produces

A an inverter
B an AND gate
C a NOR gate
D an OR gate

10 In order to be correctly connected to a microprocessor, input and output devices require

A direct memory access
B separate power supplies
C isolating transformers
D interfaces

10 Amplifiers and oscillators

Amplifiers

It was shown in the last chapter how a single transistor may be used as a switch, where it can only take up two states, either ON or OFF. We now take a look at a simple audio amplifier capable of amplifying frequencies between about 20 Hz and 20 kHz – the accepted range of audio frequencies detectable by the human ear. Figure 10.1 shows a simple audio amplifier circuit and you will probably see that it is basically the same as Figure 9.4 in the last chapter, but with some components added.

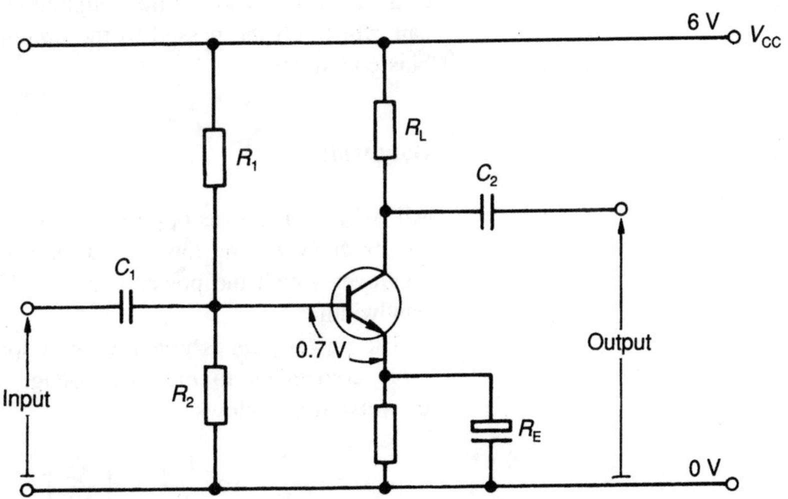

Figure 10.1 Circuit of simple audio amplifier

R_L is the load resistor while R_1 and R_2 provide base bias. R_1 and R_2 comprise a practical example of the use of a potential divider, which means that the base voltage is kept constant. This occurs because, remembering from Chapter 2, the current flowing through the potential divider is irrelevant; it's the ratio between the two resistors which determines the junction voltage.

So, if resistor values are chosen such that very much more current flows through the potential divider than flows into the transistor base, changes in base current will have a negligible effect on base voltage. An emitter resistor and capacitor have also been added to the basic circuit. The value of emitter resistor is chosen such that there is about 2 V across it in this

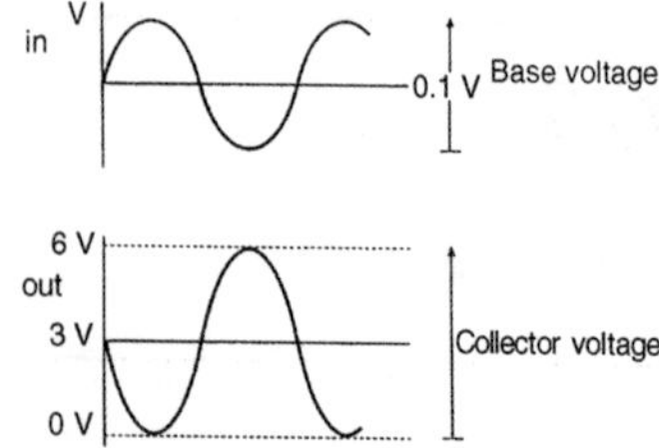

Figure 10.2 Input and output waveforms for a simple audio amplifier

type of circuit. There is normally about 0.7 V between base and emitter when the transistor is operating correctly, so there will be 2.7 V on the base itself.

When a 1 kHz sine wave signal is applied to the input, examination of the output waveform shows that it is inverted and much larger. Typical input and output waveforms are shown in Figure 10.2.

It can be seen from Figure 10.2 that as the base input signal increases, the collector voltage falls. (This was also shown in the last chapter while discussing the NOT gate or INVERTER.) This means that the signal has undergone 'phase inversion' or, to put it another way, the output signal is 180° out of phase with the input signal.

Coupling capacitors

The DC biasing components are carefully worked out to ensure correct transistor operation, so we don't want it all to be upset on the application of a signal. Inclusion of the coupling capacitors means that the AC signal can effectively be passed to the amplifier without upsetting the critical, bias conditions.

Bandwidth

All AC amplifiers (as opposed to DC amplifiers) tend to have low gain at low frequencies and low gain at high frequencies. The range of frequencies over which the power gain is half its maximum value is called the bandwidth.

The power gain is given as the quotient P_{out}/P_{in}. Because it is much more convenient to measure voltage gains, it is necessary to make a conversion as follows:

$$\text{power} = \frac{V^2}{R} \quad \text{so:} \quad \frac{P_{out}}{P_{in}} = \left(\frac{V_2}{V_1}\right)^2$$

Where V_2 is the output voltage and V_1 is the input voltage. If we want the value where the ratio of power out to power in is a half, then:

$$\frac{P_{out}}{P_{in}} = \frac{1}{2} \quad \text{and so:} \quad \frac{V_2}{V_1} = \frac{1}{\sqrt{2}}$$

1 divided by $\sqrt{2}$ is equal to 0.707, so the bandwidth is taken between the points where the voltage gain drops to 0.707 of its maximum value. It is the same as the 'half power points' which is a 3 dB drop in power gain. The bandwidth for many common audio amplifiers is 20 or 30 kHz and frequently very much higher.

In order to measure the bandwidth of an amplifier the values of V_{in} and V_{out} are obtained over a range of frequencies. The quotient V_{out}/V_{in}

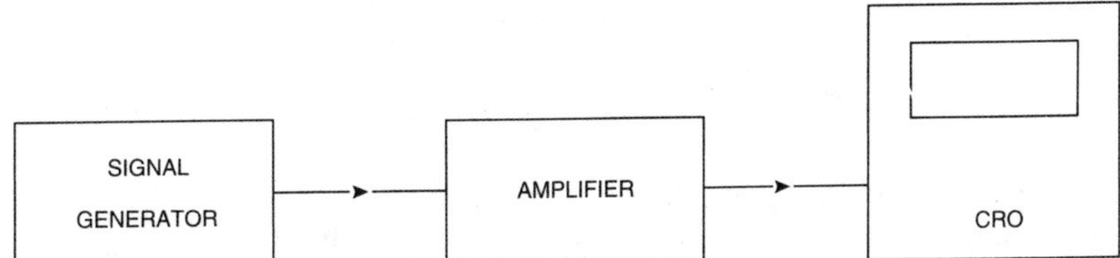

Figure 10.3 Measuring the bandwidth of an amplifier

gives the gain of the amplifier which can be plotted on a graph against the frequency. The set-up is shown in Figure 10.3, where the output is measured by a CRO. In this instance, the use of a DVM would be just as good and probably easier than using an oscilloscope. Table 10.1 lists some typical values.

Table 10.1. Typical values of V_{out} for an audio amplifier

Frequency	V_{in}	V_{out}
10 Hz	1 V	1.01 V
50 Hz	1 V	1.10 V
100 Hz	1 V	1.20 V
1 kHz	1 V	5.00 V
2 kHz	1 V	10.00 V
5 kHz	1 V	10.10 V
10 kHz	1 V	10.00 V
15 kHz	1 V	5.00 V
30 kHz	1 V	1.10 V

When the graph has been plotted, 0.707 of the maximum gain is calculated. The graph may then be used to find the two frequencies (one at each end of the graph) where the gain falls to this value. The bandwidth is then quite simply the higher frequency minus the lower one. This is shown in Figure 10.4.

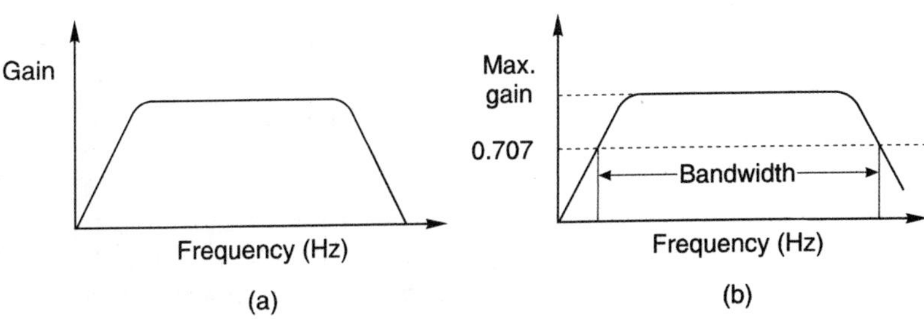

Figure 10.4 Obtaining the bandwidth of an amplifier

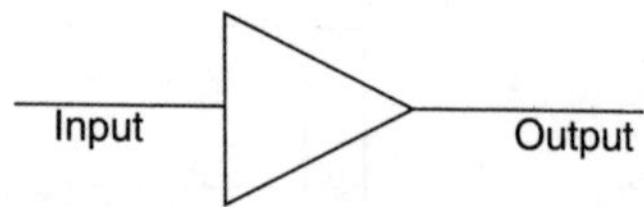

Figure 10.5 General symbol for an amplifier

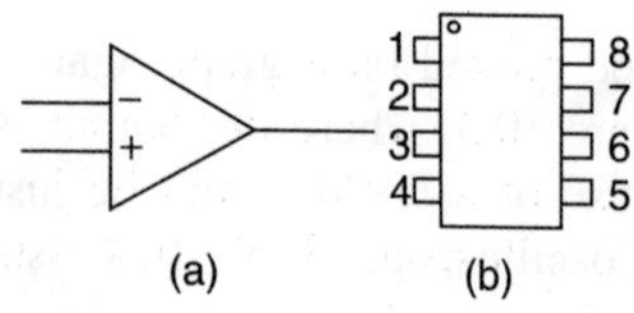

Figure 10.6 (a) Op-amp symbol. (b) 8-pin DIL package

Amplifier symbols

The general symbol for an amplifier is shown in Figure 10.5.

Operational amplifiers

Operational amplifiers (op-amps) were originally made from discrete components. They were designed to solve mathematical equations electronically, by performing such operations as addition and subtraction and so on, in analogue computers. Nowadays in IC form they have many uses, one of the most important being as high-gain DC and AC voltage amplifiers. A typical op-amp like the 741 contains about 20 transistors as well as associated resistors and capacitors, all fitted into a tiny, 8-pin DIL package. Op-amps tend to have two inputs marked '−' and '+', being the inverting and non-inverting inputs respectively. The symbol and package are shown in Figures 10.6(a) and (b).

Audio amplifiers

A typical AF (audio frequency) amplifier is shown in Figure 10.7. One application of such an amplifier is in the first AF stage of a television receiver, but similar amplifiers are also used in hi-fi units and radios and as interstage amplifiers in other audio equipment.

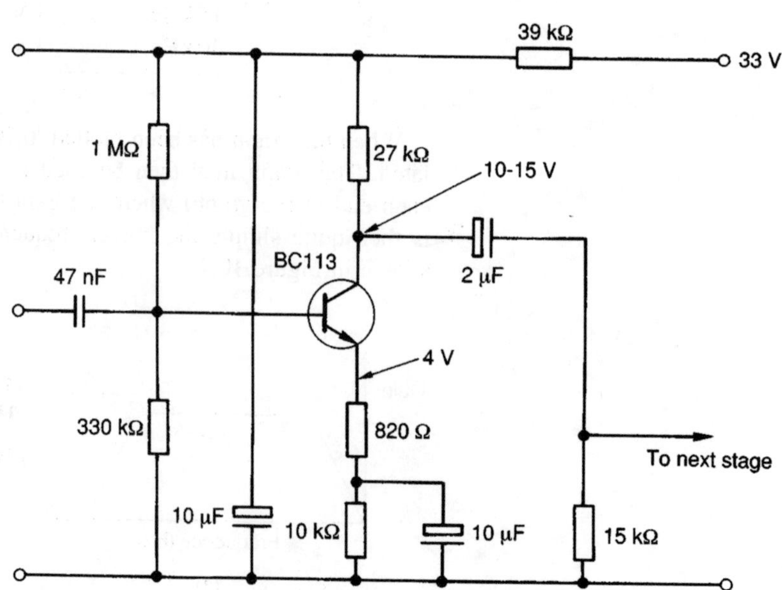

Figure 10.7 Typical audio amplifier in a TV receiver

The amplifier will draw some current even though no signal is applied. This is called the 'quiescent' current and can be rather high; however, this is not a problem in low power signal circuits. The transistor is connected in common emitter mode which means that the input signal, the output signal and the emitter of the transistor are all connected to a common point.

This type of 'Class A' amplifier gives no distortion and provides high current and voltage amplification. The input and output impedances are 'medium', being of the order of a few kΩ. The output is phase inverted compared to the input as we have already seen.

Figure 10.8 gives an example of an amplifier connected in common base configuration. There is no signal inversion, the current amplification is less than 1, but there is a large voltage amplification. This mode provides a good signal to noise ratio and is not prone to instability. One common application is in VHF radio amplifiers, but in this case, the circuit is designed to amplify the output from a low impedance microphone.

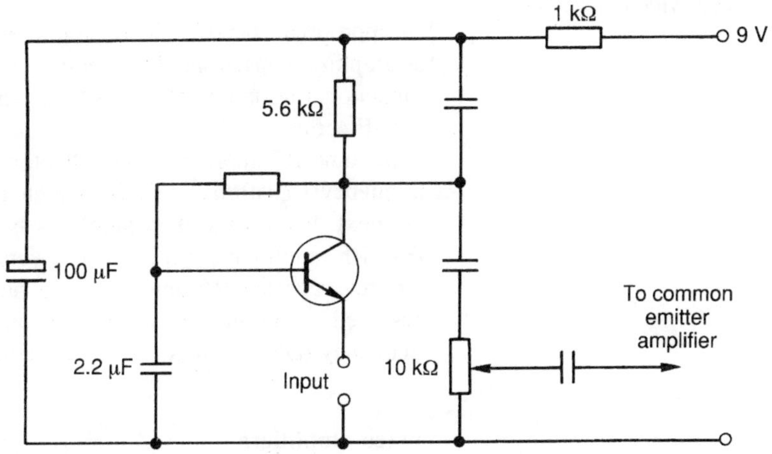

Figure 10.8 A low impedance microphone amplifier

An example of the common collector configuration is given in Figure 10.9. These designs are more commonly known as emitter followers as the emitter voltage 'follows' the base voltage, less the 0.7 V base/emitter drop. Such circuits are characterised by having a voltage gain of less than 1, a very large current gain and a low impedance output. This makes them very suitable for driving loads such as motors, loudspeakers or tone control circuits. There is no signal inversion.

IF and RF amplifiers

Intermediate frequency (IF) amplifiers are designed to work on single frequencies, rather than a band of frequencies, although they must have

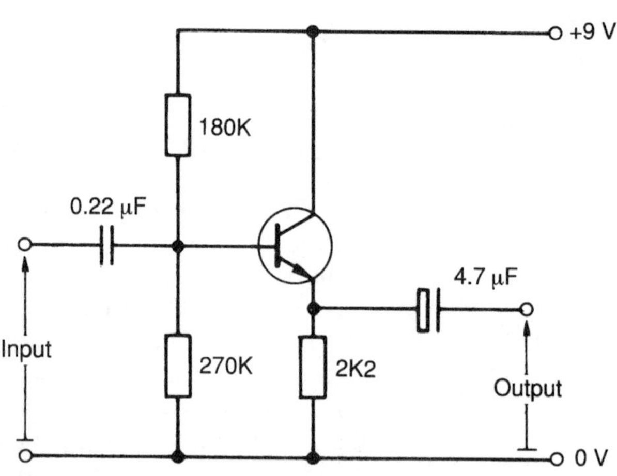

Figure 10.9 An emitter follower amplifier

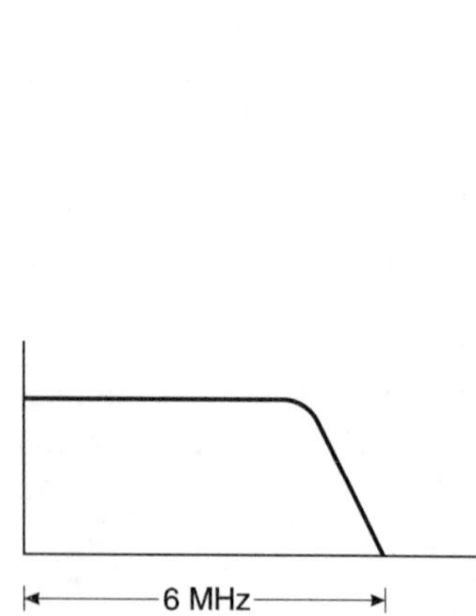

Figure 10.10 Bandwidth of a typical video amplifier

an appropriate bandwidth. IF amplifiers tend to have a large amount of decoupling (capacitors placed across the power lines), low value coupling capacitors (around 80 pF to 1000 pF) and a low value of emitter capacitor (10 nF perhaps).

In some IF amplifiers, the collector loads may consist of intermediate frequency transformers (IFTs) which pass the signal from one stage to the next. Each stage then amplifies best at the natural frequency of the IFT. RF amplifiers tend to have a very similar design; in fact, the main difference is that RF amplifiers are tunable. Radio receivers are usually designed so that most of the amplification occurs at the fixed intermediate frequency rather than at a variable radio frequency.

Video amplifiers

These amplifiers, used for the video circuits in television receivers, must be able to amplify all frequencies from zero up to about 6 MHz. In Chapter 8 we saw that the UK standard of television transmission has 625 lines on the screen with a frame rate of 25 Hz producing a line frequency of 15 625 Hz. If there are about 400 variations in picture brightness along each line, then the video signal applied to the TV tube (often the cathode) must change in amplitude this number of times per second, which is $15\,625 \times 400 =$ 6.25 MHz. Hence, any video amplifier must be capable of handling all the frequencies from zero up to this approximate value.

Because of the wide range of frequencies video amplifiers have to handle, they are also known as wide-band amplifiers. Such amplifiers are also used as Y amplifiers in oscilloscopes and have many applications in radar equipment and digital data signals. The bandwidth of a typical TV video amplifier would look something like that shown in Figure 10.10.

Operational amplifiers

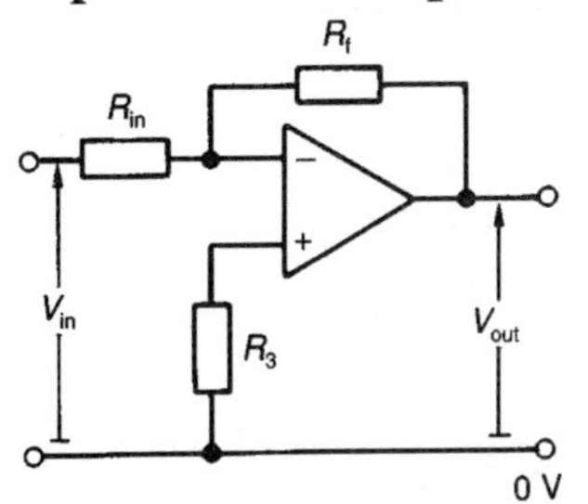

Figure 10.11 An op-amp inverting amplifier

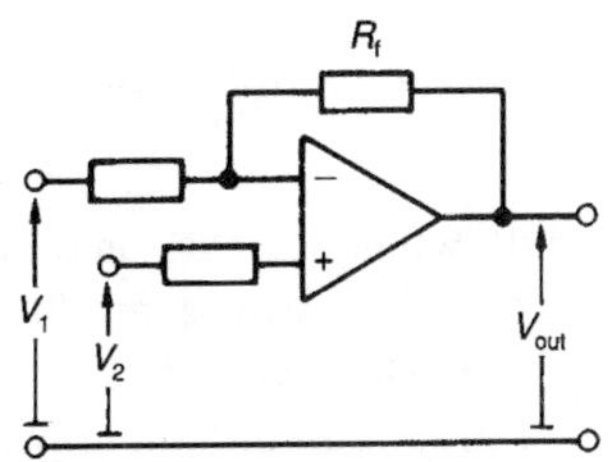

Figure 10.12 An op-amp differential amplifier

These devices were introduced earlier in this chapter and Figure 10.6 shows the symbol and package for the popular 741 op-amp. The 741 can be operated as a normal inverting amplifier by connecting it as shown in Figure 10.11. The gain of the amplifier is entirely dependent upon the value of the input resistor (R_{in}) and feedback resistor (R_f). Without any negative feedback, the 741 has a gain of about 10^5 and when it is used in this way is rather unstable in most applications.

R_3 tends to have the same value as R_{in}. The gain of the amplifier, V_{out}/V_{in}, is equal to $-R_f/R_{in}$. The negative sign indicates that the amplifier is inverting.

Finally, on the subject of op-amps, Figure 10.12 shows a differential amplifier. This configuration was mentioned in Chapter 8 where a differential amplifier is illustrated in Figure 8.2. The output from this amplifier depends upon the difference between the two inputs and not the actual values themselves. The gain of the amplifier, however, is still governed by the amount of feedback and is given by:

$$\frac{V_{out}}{V_2 - V_1} = \frac{R_f}{R_{in}}$$

Difference amplifiers are often configured to have a gain of unity so that the output is simply the difference between the two inputs. If V_1 is greater than V_2 then the output is negative; if V_2 is greater than V_1 then the output is positive.

Oscillators

Since a single-stage transistor amplifier operating in common emitter mode produces an output which is 180° out of phase with its input, any portion of the signal fed back to the input will be negative feedback. This has the effect of reducing the gain of the amplifier, widening the bandwidth and increasing stability.

If part of the output from an amplifier is fed back in phase with the input, then the amplifier will tend to oscillate. Hence, an oscillator is simply an amplifier with its output connected back to its input, in phase (or 360° out of phase), with sufficient gain for it to overcome any losses in the system.

Figure 10.13 shows two simple amplifiers; the output of the first is connected to the input of the second, and the output of the second is connected back to the input of the first. If the circuit is redrawn it becomes the familiar astable multivibrator configuration we saw in the last chapter (Figure 9.17).

Since an ordinary one-transistor amplifier in common emitter mode produces a 180° phase change (inversion), if a further 180° phase change is produced by other means, then the circuit will oscillate. This is shown in Figure 10.14.

A differentiating network is shown and described in Chapter 3 (Figure 3.24). If the network is fed with a sine wave, rather than a square wave, then the output is identical to the input but shifted in phase. If there

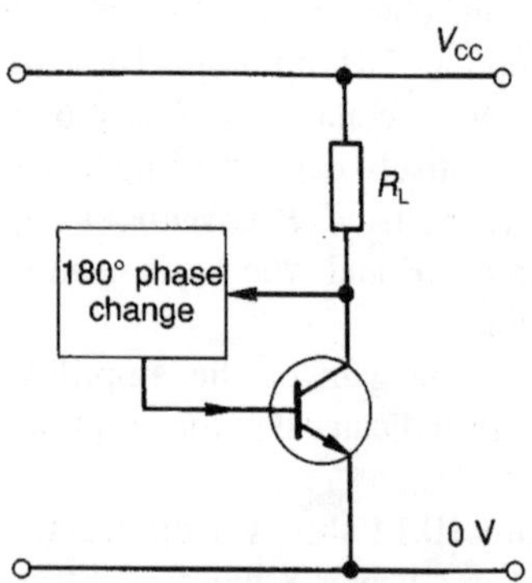

Figure 10.14 The transistor produces 180° phase shift; a further 180° in the feedback loop produces a total of 360° and the circuit oscillates

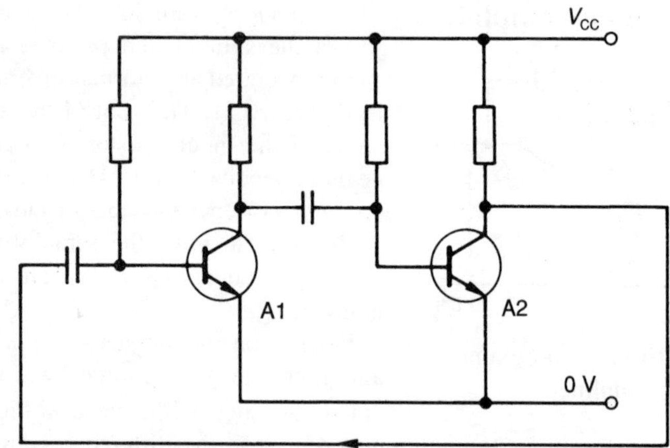

Figure 10.13 Two amplifiers back to back producing an oscillator

were no resistance in the circuit, the phase shift would be 90°, and so two such circuits in the feedback loop of an amplifier would produce a 180° phase shift and the circuit would oscillate. However, it is only possible to produce 90° if there's no resistance in the circuit and there will always be some, even if it's only the resistance of the wires connecting the circuit. As the circuit resistance is increased, the phase shift is reduced, so, by careful calculation, the phase shift can be reduced to 60° and three such circuits in the feedback loop will produce the 180° phase shift required.

Thus, the transistor produces a 180° phase shift and the feedback network produces another 180° (360° in all), and the circuit oscillates, producing a sine wave. The result is a phase shift oscillator and a typical circuit is shown in Figure 10.15.

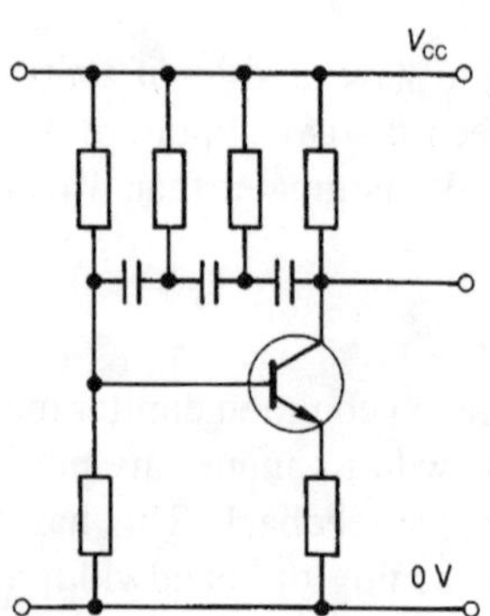

Figure 10.15 The phase shift oscillator

There are many other designs for sine wave oscillators but they all fall into two main groups: *RC* and *LC*. The former are resistance/capacitance oscillators (like the phase shift oscillator, or 'ladder' network), usually used for frequencies less than 50 kHz; the latter are inductance/capacitance oscillators, usually used for frequencies greater than 50 kHz. Other oscillator groupings and types are shown in Table 10.2.

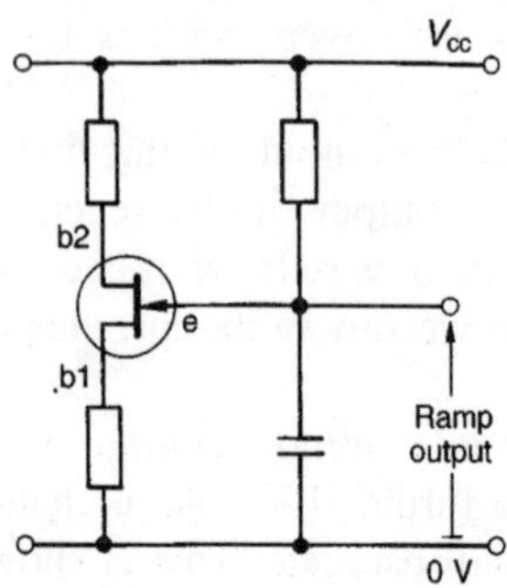

Figure 10.16 A UJT circuit which produces a ramp waveform

Ramp or sawtooth oscillators

Generally, the words 'ramp' and 'sawtooth' mean much the same thing when describing this kind of oscillator. They have many applications in the fields of computers, communication and automatic control systems. Ramp waveforms may be easily generated with the simple circuit of Figure 10.16 which uses a unijunction transistor (UJT).

All ramp waveforms have a common characteristic, whether the pulses are widely or closely spaced in time, have two ramp edges, or a single,

Table 10.2. One grouping for the many types of oscillator. There are basically three distinct types and some sub-classifications are given. In the square wave section 'M/V' stands for multivibrator

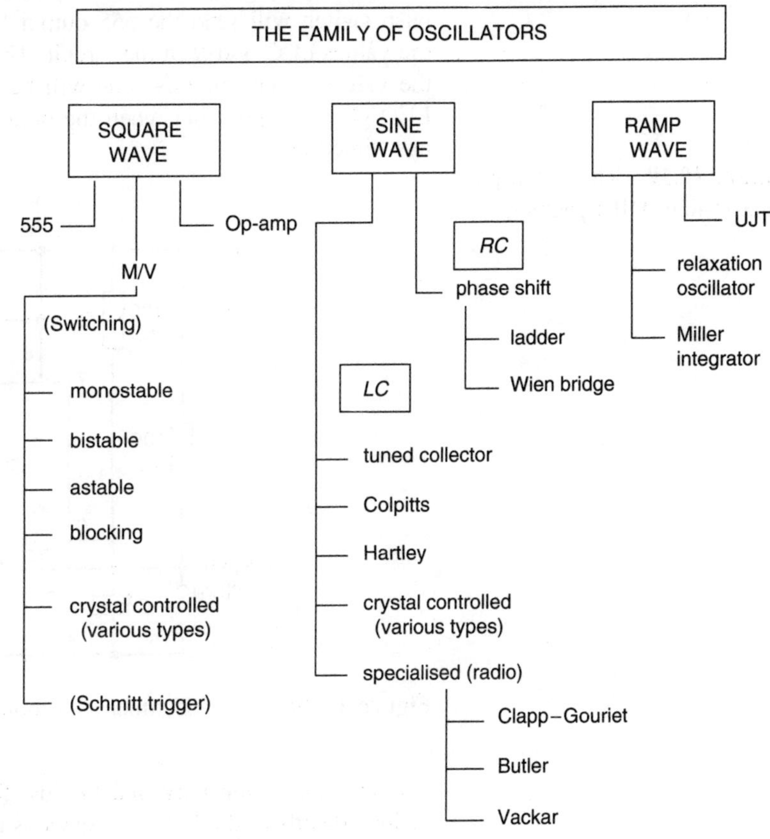

(a) (b)

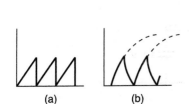

Figure 10.17 (a) Idealised ramp waveform. (b) Practical ramp waveform, using the initial exponential rise of a charging capacitor

linear leading edge. Every ramp waveform has at least one edge, usually the leading one, which is approximately a linear function of time. In fact, a ramp waveform is essentially an exponential with a large time constant compared to the time duration of the ramp. An idealised ramp waveform is shown in Figure 10.17(a). The increase in voltage of the ramp is linear – that is to say, when the same time elapses, the same increase in voltage is obtained. The fast rising initial exponential charge of a capacitor may be used as shown in Figure 10.17(b).

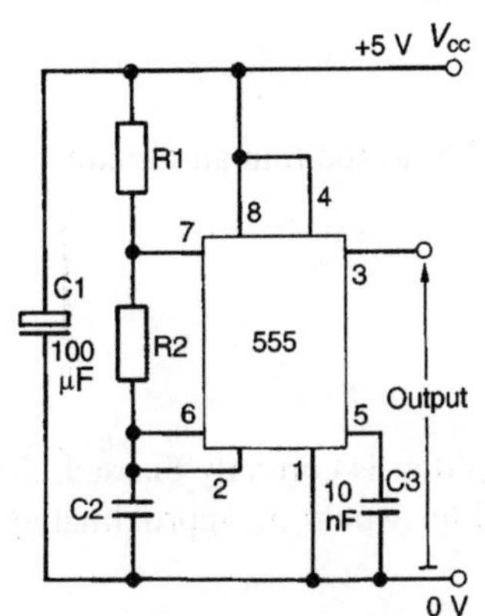

Figure 10.18 A 555 timer connected in astable mode

555 oscillators

The 555 'timer' IC may be used to produce a square wave, but it also generates a ramp waveform at pin 6 when connected in astable mode. A typical circuit is shown in Figure 10.18.

This circuit will produce an audible tone when a loudspeaker is connected to its output. Pin-out details are given in Figure 10.19.

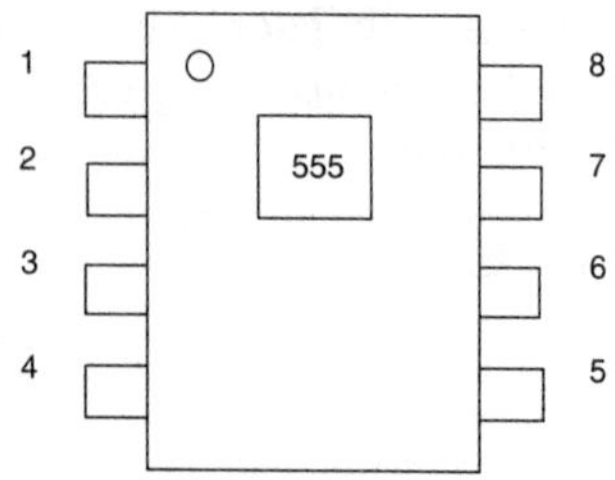

Figure 10.19 The 555 8-pin dual-in-line (DIL) package

A monostable using a 555 IC

The circuit of Figure 10.20 shows a 555 in monostable mode. Pressing the push switch will send the 555 output 'high' for a period determined by the values of C and R in the circuit. The time constant, $\tau = 1.1CR$. With the values given, the ON time will be about 5 s. In its resting state, the LED glows; it goes out when the button is pressed until the monostable has 'timed out'.

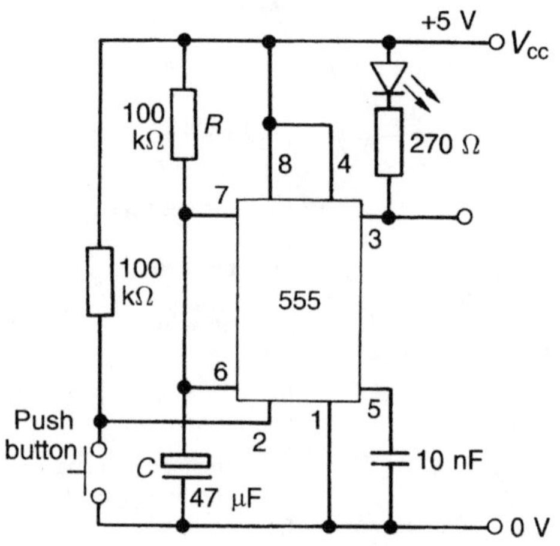

Figure 10.20 The 555 timer as a monostable

Even though the City and Guilds 224 syllabus does not specifically include details of the 555, the device is often used in the practical test and is so commonly used in electronics that it is useful now to have the basic knowledge outlined in this chapter. Further details are given in Chapter 19, Assignment 4.

Multiple choice questions

1 The accepted range of frequencies audible to the human ear are

A 300 Hz to 3 KHz
B 20 Hz to 10 kHz
C 20 Hz to 20 kHz
D 100 Hz to 100 kHz

2 When a transistor in a simple audio amplifier is correctly biassed, the expected voltage between base and emitter would be approximately

A 100 mV
B 700 mV
C 0.5 V
D 2.0 V

3 The main advantage of Class A amplifier operation is

A an undistorted output is obtained
B quiescent current is negligible
C the output power is extremely high
D the output impedance is extremely low

4 Thermal runaway may occur in a transistor because

A it is made from a semiconductor material
B the current supplied to the device is too high
C the power supply connections are reversed
D an emitter capacitor has not been fitted

5 A simple one transistor amplifier is operating in Class A mode. A potential divider is chosen to provide base bias so that

A the current is limited to a safe level
B the resistor values may be more easily calculated
C the current can be set at a level such that changes in base current have a negligible effect on base bias voltage
D negative feedback can be eliminated

6 A simple method of calculating the bandwidth of an amplifier is to obtain various values of

A V_{in}/V_{out}
B V_{out}/V_{in}
C $V_{out} - V_{in}$
D $V_{out} + V_{in}$

7 Another name for a transistor connected as an emitter follower is

A common emitter
B common base
C common collector
D super-alpha

8 The bandwidth of a video amplifier is typically

A 20 kHz
B 100 kHz
C 6 MHz
D 10.7 MHz

9 In a one-transistor phase shift oscillator, the phase shift required by the feedback circuit must be

A 90°
B 180°
C 270°
D 360°

10 As a rough guide, *RC* and *LC* oscillators are generally used for frequencies

	RC	*LC*
A	<50 kHz	>50 kHz
B	<50 kHz	>6 MHz
C	<6 MHz	>10 MHz
D	<10 MHz	>10 MHz

11 A ramp waveform is more or less the same as a

A sine wave
B square wave
C sawtooth wave
D pulse waveform

12 Connected in the astable configuration, a 555 timer can provide

A sine and ramp waveforms
B square and ramp waveforms
C sine and square waveforms
D pulse waveforms

13 The term 'flip-flop' is another name for

A an oscillator
B an astable
C a monostable
D a bistable

14 Reducing the gain of an amplifier by using negative feedback increases the amplifier's

A bandwidth
B sensitivity
C distortion
D current consumption

15 When a sine wave is applied to a single differentiating network the maximum phase shift obtainable is

A 60°
B 90°
C 180°
D 270°

11 Filters and waveshaping

Introduction

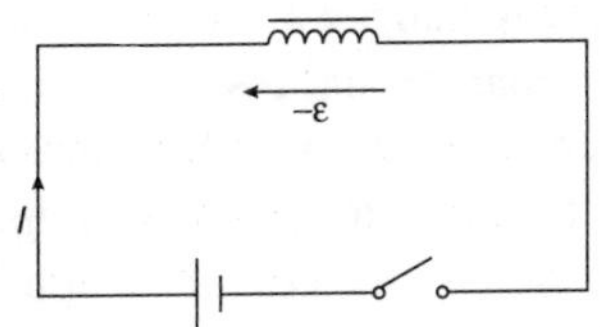

Figure 11.1 Production of a back emf when the current to a coil is first applied, or interrupted

We saw in Chapter 4 that a solenoid or coil of wire carrying a current produces a magnetic field. We also noted that a changing magnetic field in the vicinity of a coil will induce an emf in that coil – the basic principle of electromagnetic induction. It is this important effect that we return to here.

If a coil carries a current as shown in Figure 11.1, it will produce a magnetic field. If the current to the coil is interrupted, the magnetic field collapses. It follows that the collapsing magnetic field is a *changing* magnetic field which must, therefore, induce an emf in the coil. From Lenz's law, the induced current tries to oppose the 'motion' causing it and therefore acts in the opposite direction to the main current which set up the magnetic field in the first place. The emf is therefore called a 'back emf'.

If an alternating current is applied across the coil, then a constant back emf is produced which opposes the main current. This is because the AC source is a constantly varying voltage and current, producing a constantly varying magnetic field, constantly inducing an emf in the coil. So the coil is in fact an electronic component which opposes the flow of current in an AC circuit.

As far as DC is concerned, the coil only opposes current because of its resistance, generally a very small amount. In an AC circuit, the back emf produces an additional opposition to current flow. Since this opposition is quite a different mechanism to electrical resistance, it is given a different name – reactance. Reactance is given the symbol X and, like resistance, is measured in ohms (Ω).

The general term for a coil is an inductor, because it produces the effect of self-induction which has been described before. In the same way that we talk about a resistor having resistance, an inductor has inductance, a quantity which has the unit of Henrys, named in honour of Joseph Henry, an American scientist who independently discovered electromagnetic induction round about the same time (1831) as Michael Faraday in England.

The symbol for inductance is L; although the unit henry (H) is quite large, coils of 1 H (and larger) may be found in some power supplies, but in radio work, mH and μH are more common. Some inductors are called 'chokes' as they offer a large opposition (reactance) to alternating current. The circuit symbols are shown in Chapter 4 (Figure 4.28) where other basic information on this topic is also provided.

Remembering the information in Chapter 4, it was noted that the emf induced in a coil depends upon various factors. In particular, the speed with which the magnetic field changes affects the emf (and back emf where appropriate); the greater the speed, the greater the back emf. It follows, therefore, that a given inductor will have an opposition to an alternating current which increases with the frequency of the supply. Hence, we have a component which is sensitive to frequency; low frequencies passing through the inductor with comparative ease, higher frequencies being attenuated.

Leaving inductors aside for one moment, let's take a look at how a capacitor reacts to an alternating current. This was examined in the last chapter where we noted that, from a simplistic view, a capacitor will pass an AC signal, but block DC, because a DC signal charges a capacitor and then nothing occurs. An AC signal constantly charges and discharges the capacitor, giving the effect of a constant current flowing in such a circuit. Furthermore, the greater the speed at which the capacitor is charged and discharged, the greater the flow of current in the circuit. Let's now think about what happens in a circuit like that of Figure 11.2.

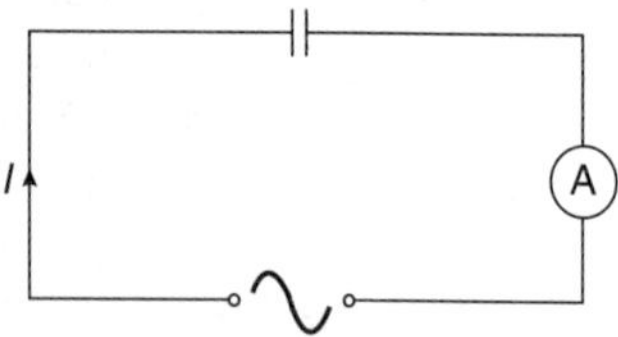

Figure 11.2 A capacitor in an AC circuit

When the AC supply is connected, the ammeter will register a current which is a function of the size of the capacitor and the frequency of the supply. As the frequency is increased, so too is the current, and it looks as though the opposition to the supply current reduces – the current in the circuit increases. If the current increases then something like 'resistance' is being reduced. However, there is no resistance in the circuit (apart from the connecting wires, where it is so small we can ignore it), so what is it?

The answer is that the capacitor also has reactance, the same quantity we identified when an inductor is supplied with an AC signal. However, unlike the inductor where low frequencies may pass with comparative ease and high frequencies are attenuated, the capacitor has the opposite effect, higher frequencies 'passing' through with comparative ease and lower frequencies being attenuated.

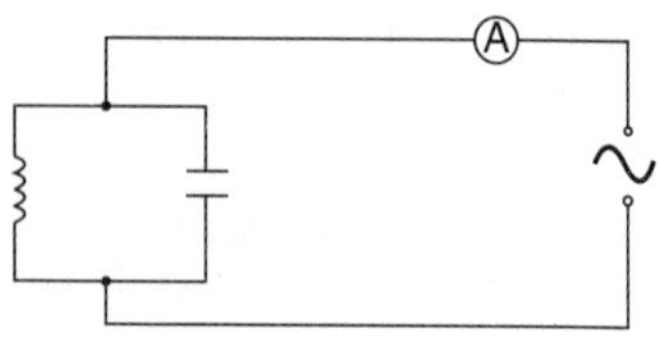

Figure 11.3 An inductor and capacitor wired in parallel

There is an extremely important conclusion to consider here: if the two components are connected in parallel as shown in Figure 11.3 then, starting with a low frequency and gradually increasing, the inductor allows a large current to flow and the capacitor blocks the current. As the frequency is increased further, the capacitor begins to 'conduct' more while the inductor blocks the current.

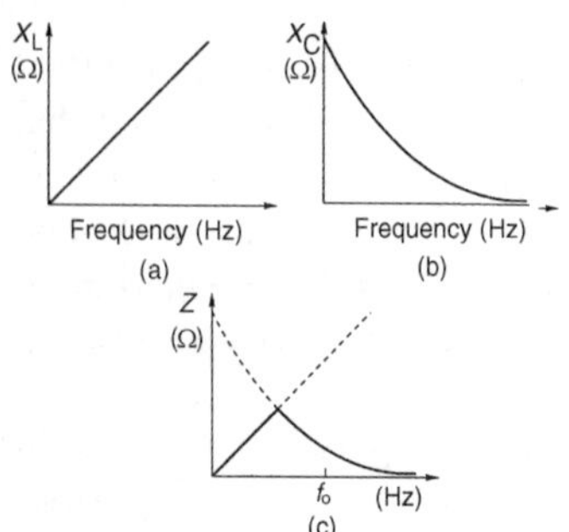

Figure 11.4 (a) Inductive reactance (X_L) increases with frequency. (b) Capacitive reactance decreases with frequency. (c) Impedance rises to a maximum at the resonant frequency (f_o)

If these effects are plotted separately on a graph we should obtain results similar to Figures 11.4(a) and 11.4(b). Combining the two together would produce something like Figure 11.4(c) showing that the circuit has its maximum reactance at one frequency, called the resonant frequency, f_o. It will be noticed that in Figure 11.4(c), the symbol 'Z' is used. This is because the combination of capacitive reactance, inductive reactance and any resistance in the circuit is called impedance, symbol Z, units also in ohms (Ω).

As you will no doubt be aware, there is much more to this subject, but, we have already looked into the matter more deeply than the City and Guilds 224 Part 1 syllabus requires. However, it is felt that it makes more sense to study these elementary ideas in order that a greater understanding of the design and operation of filters be more easily understood and we move onto that subject now.

Filters

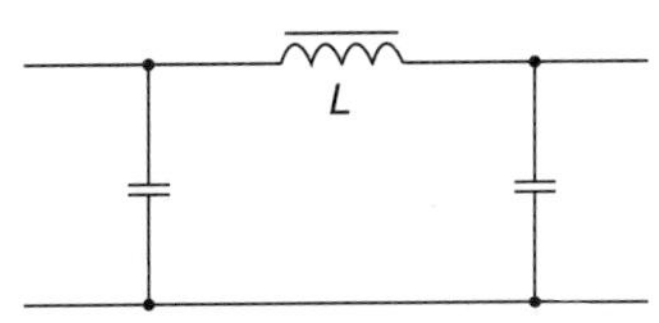

Figure 11.5 A pi filter

In Chapter 6, we discussed the various elements of a power supply unit (PSU), and evolved the idea of a pi filter. It is useful now to return to this and examine the filter in terms of the associated reactances. A typical pi filter is reproduced in Figure 11.5.

You will recall that by the time the mains arrives at the pi filter in a PSU, it has been transformed down to a much lower voltage, say 12 V, and rectified so that only the positive-going parts of the sine wave make it to this point. There is still some 'ripple', however, and this takes the form of a varying current and voltage. We now see that this varying voltage will charge and discharge the capacitors which hence present a low 'resistance' (reactance, correctly speaking) down to the 0 V rail. At the same time, the inductor presents a very large reactance so that any ripple 'sees' a very large reactance. The overall effect is to allow the DC component to flow easily to the load and the 50 Hz (or 100 Hz for full wave rectification) ripple is successfully eliminated.

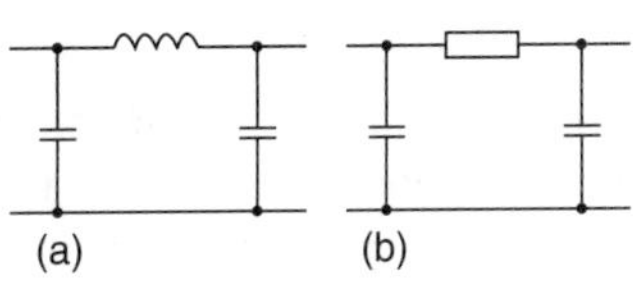

Figure 11.6 Two variations of low pass filters

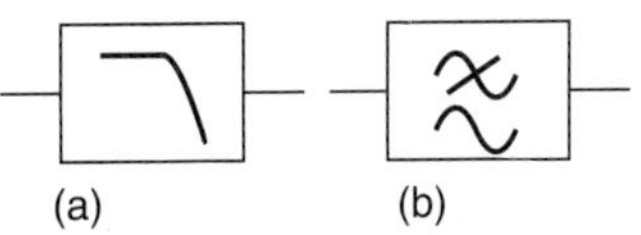

Figure 11.7 Symbols used for low pass filters

The use of an inductor as part of the smoothing process in a PSU is less common now than it used to be. In the good old days of valve amplifiers, when anode voltages may rise to 600 V or even more, inductive smoothing was very effective. However, now that transistor amplifiers dominate, inductors are used less. Because transistors tend to be used at low voltages and higher currents, the associated inductors must be wound with thicker wire and there is always the tendency for them to saturate – that is to say, reach a maximum point of effectiveness. For these reasons (and some others, like cost and radiation) resistors are used instead.

The pi filter is actually a 'low pass filter' (LPF) because the arrangement presents a very low impedance to low frequencies and passes the most amount of current when presented with a DC signal. Filters like this (with quite different component values from those quoted for their use in power supplies) are used in radio transmitters (particularly amateur ones) so that the required signal is passed to the antenna while unwanted, high frequency harmonics are suppressed. You may find it instructive to obtain a copy of the booklet *How to improve radio and TV reception*, which contains a lot of technical information on this subject and is free from most main Post Offices.

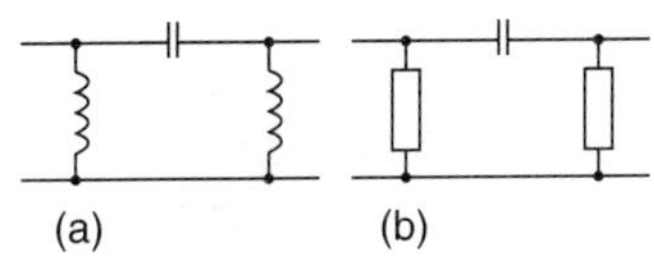

Figure 11.8 Two variations of high pass filters

Two variations of a typical low pass filter are shown in Figure 11.6, and two variations of symbols are given in Figure 11.7. Either symbol may be use for either configuration.

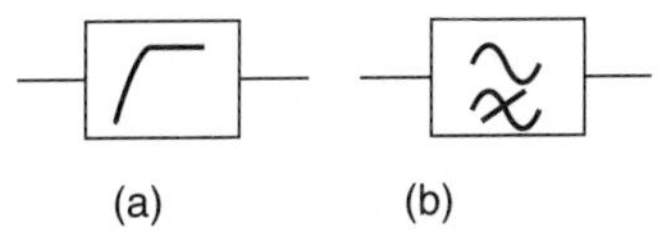

Figure 11.9 Symbols used for high pass filters

If the components are interchanged as shown in Figure 11.8, a high pass filter is produced. Remembering the theory outlined previously, the higher the frequency, the better the capacitor 'conducts' and the less the inductor conducts. High frequencies therefore find a hard path through the inductor and an easy one through the capacitor.

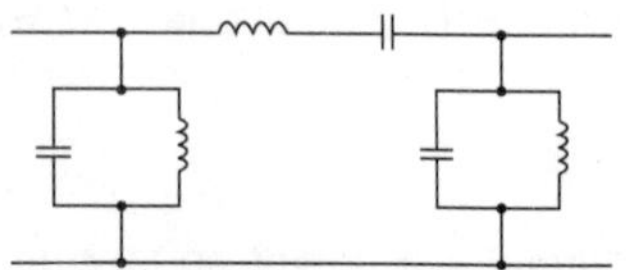

Figure 11.10 A typical band pass filter. This permits a band of frequencies to pass through it

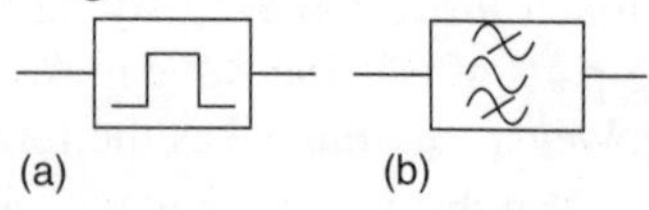

Figure 11.11 Symbols used for band pass filters

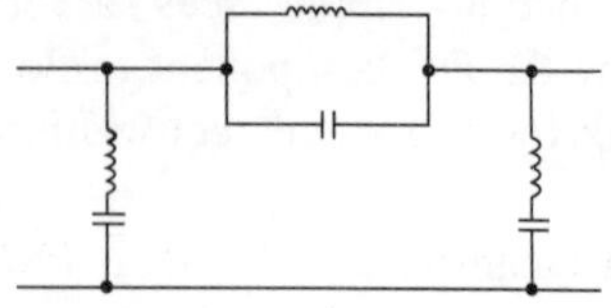

Figure 11.12 A typical band stop filter

The upshot of this is that high frequencies pass more easily through the high pass filter (HPF) and lower frequencies are attenuated. The equivalent symbols are shown in Figure 11.9.

Band pass filters

These filters make use of series and parallel tuned circuits in order to pass a 'band' of frequencies. Figure 11.10 shows a typical configuration; the appropriate symbols are given in Figure 11.11. It has already been shown that a parallel tuned circuit has maximum impedance at the resonant frequency. The filter is therefore designed to have a resonant frequency which is in the centre of the band of frequencies being selected. Conversely, the series tuned circuit has minimum impedance at the centre frequency.

Band stop filters

As the name implies these filters are designed to stop a range of frequencies. Frequencies which are much lower or much higher than the centre frequency are attenuated. A typical band stop filter is shown in Figure 11.12 and the symbols in Figure 11.13.

Clipping, restoring and limiting

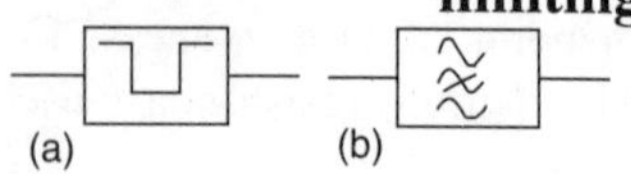

Figure 11.13 Symbols used for band stop filters

1 CLIPPING removes part of a waveform; either the positive-going part only, the negative part only, or both parts.
2 RESTORING – or clamping – provides a signal with a DC level that may have been lost previously because of capacitive coupling.
3 LIMITING limits an AC signal to a maximum amplitude. One important example is in FM detection, to ensure that the signal has a constant amplitude.

1 Clippers

The use of sine waves is assumed in these notes, but clippers and clampers work equally well with other waveforms. The simplest clipper is a half wave rectifier as shown in Figure 11.14(a), but the configuration shown in Figure 11.14(b) produces the same effect. Reversing the diode allows the negative-going half-cycles to appear at the output.

Other clipping circuits have the effects shown in Figure 11.15. In the case of clipper 3, a sine wave may be converted into an approximate square wave.

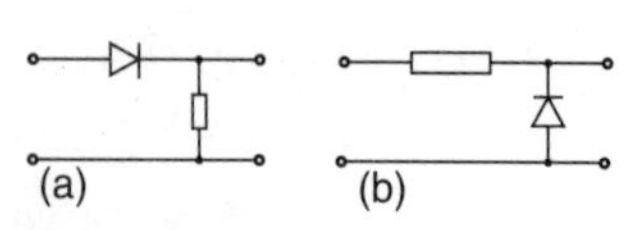

Figure 11.14 (a) The simplest clipper is a half wave rectifier. (b) Another configuration has the same effect

2 Clamping or restoring

Whenever a signal is passed through a capacitor, any DC level present in the original waveform is lost. A diode may be used to restore the DC

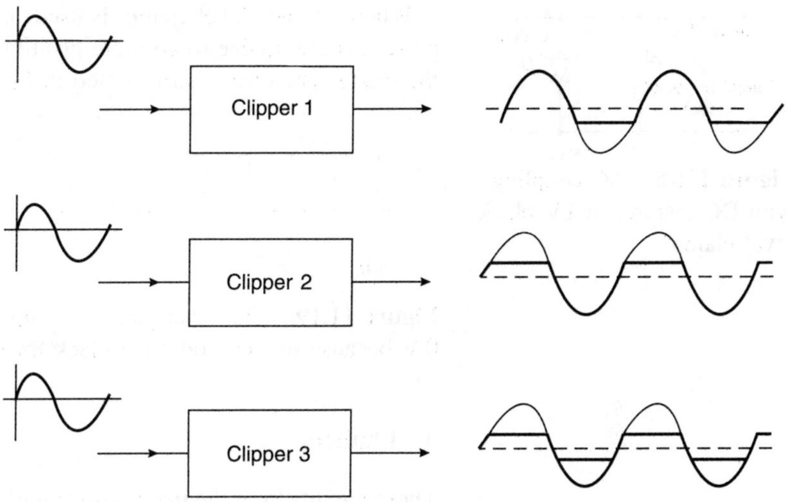

Clipper 1

Clipper 2

Clipper 3

Figure 11.15 Effect of other clipping circuits

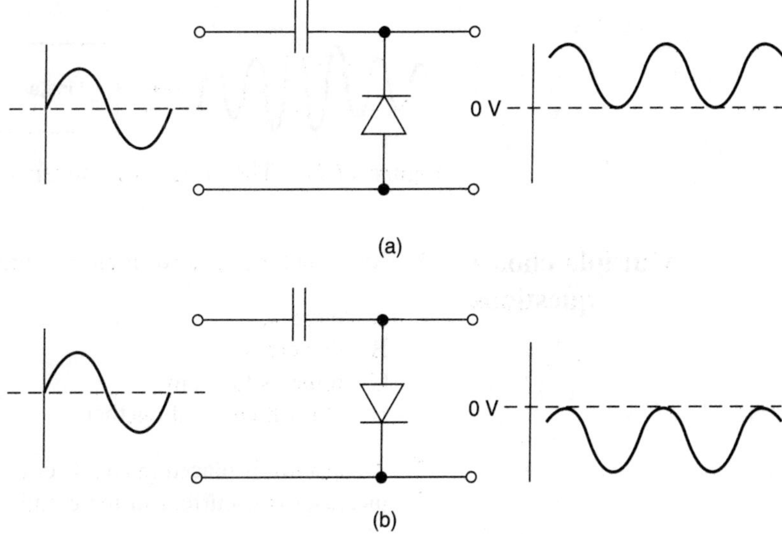

Figure 11.16 Two simple examples of DC level restoration

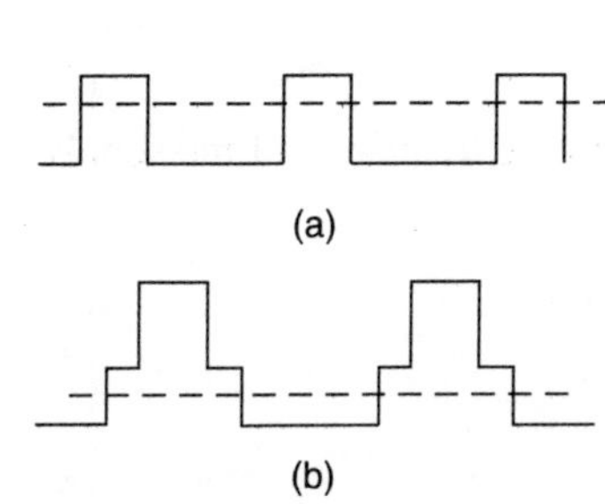

Figure 11.17 Without a black level clamp, the black level floats about. (a) A dark picture; the black level is below mean level. (b) A bright picture; the black level is now above mean level

level and clamp it at that value. Either the positive or negative peak value of a waveform may be clamped to some desired level. Two simple DC restorer circuits are shown in Figures 11.16(a) and (b).

In a television receiver the video waveform is clamped at the pedestal level thus effectively restoring the black level established in the studio. Hence, in this example, the DC restorer determines the correct signal brightness level at the receiver. Without the clamper, the signal black level floats about as the brightness content of the TV signal varies. This is shown in Figure 11.17.

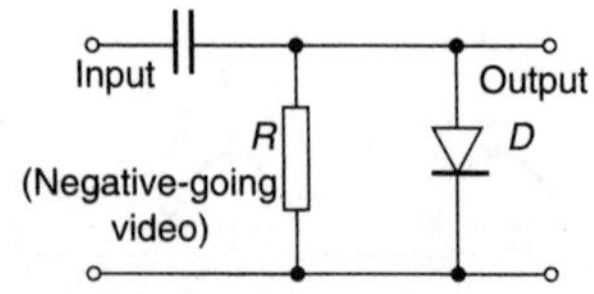

Figure 11.18 AC coupling with DC restorer; a TV black level clamp

When a 'black level clamp' is used, as shown in Figure 11.18, the sync pulse tips are unable to go more positive than 0 V (zero volts) because of the diode. The effect is illustrated in Figure 11.19.

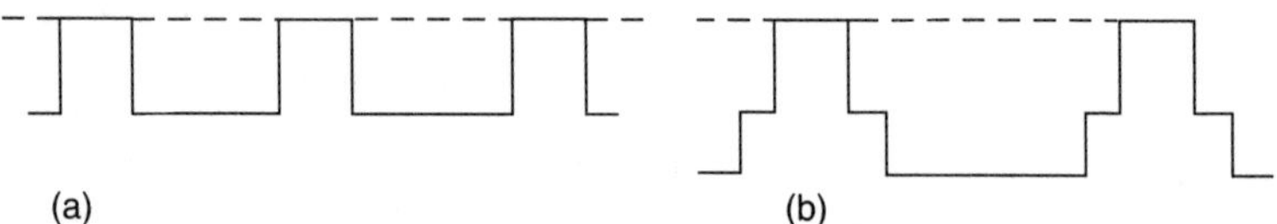

Figure 11.19 The sync pulse tips are unable to go more positive than 0 V because of the diode; the black level is held constant

3 Limiters

These circuits limit the maximum amplitude of a signal. The most common example of their use is in FM radio receiver detector circuits. The effect is shown in Figure 11.20.

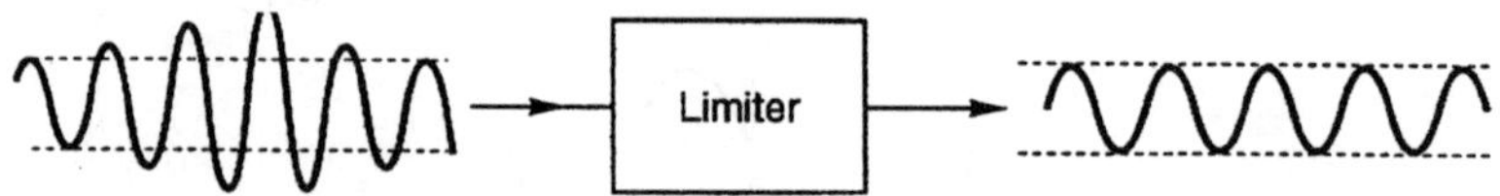

Figure 11.20 The effect of a limiter on a signal of varying amplitude

Multiple choice questions

1 An inductor is placed in an AC circuit. The current in the circuit

A increases
B decreases
C remains the same
D stops flowing altogether

2 A capacitor is placed in an AC circuit. As the frequency of the supply increases, the current in the circuit

A reduces
B increases
C remains the same
D produces a back emf

3 The type of filter most commonly used in a power supply for smoothing is

A a pi filter
B a high pass filter
C a band pass filter
D a band stop filter

4 A clipping circuit

A removes only the positive-going part of a waveform
B removes only the negative-going part of a waveform
C may remove both positive- and negative-going parts of a waveform
D has no effect on AC signals

5 A DC restorer

A converts AC into DC
B converts DC into AC
C provides a signal with a DC level
D removes the DC level of a signal

6 A television 'black level clamp' is an example of

A a high pass filter
B a low pass filter
C a clipper
D a DC restorer

7 A limiting circuit may be used for

A reducing negative feedback in an amplifier
B reducing positive feedback in an oscillator
C establishing a lost DC level
D maintaining the maximum amplitude of a signal

8 The unit of ohms is applied to

A resistance only
B reactance and resistance only
C reactance and impedance only
D resistance, reactance and impedance

9 A filter which allows a DC current to pass through it most easily may be called

A a low pass filter
B a high pass filter
C a band pass filter
D a band stop filter

10 The name of the circuit which can change a sine wave into an approximate square wave is

A a DC restorer
B an AC restorer
C a limiter
D a clipper

12 Transducers

Introduction

A transducer is generally regarded as being a device which converts one form of energy into another. A typical, example is a loudspeaker. This is a transducer whose input consists of a varying electric waveform and whose output is an acoustic waveform, or, putting it simply, sound.

More technically, a transducer may be described as a device by means of which energy can flow from one transmission system to another. The energy transmitted by these systems may be of any form, electrical, mechanical or acoustical, and it may be of the same form or of different forms in the various input and output systems.

In the example given above (the loudspeaker), the input is electrical and the output is acoustical; the input and output energy forms are different. In the case of a transformer, the input voltage is changed to a lower or higher output voltage; in this case the input and output energy forms are the same.

Table 12.1 shows some more common examples of transducers.

Aerial and antenna

The convention adopted in Table 12.1 is that an antenna is used for radio transmission and an aerial is used for reception, although in practice the two terms are generally interchangeable. Since detailed knowledge of many of the transducers listed in Table 12.1 is not required, they can all be represented by a small box showing the type of energy input and the type of energy output. The antenna and aerial are used to demonstrate this as shown in Figure 12.1.

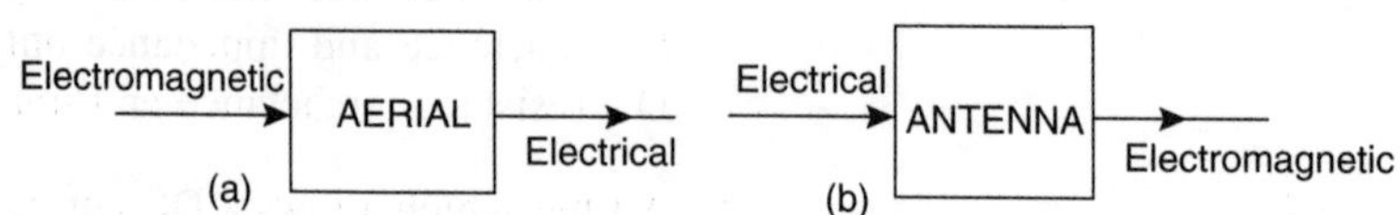

Figure 12.1 The aerial (a) and antenna (b), as transducers

The same format may be used for any of the transducers listed, so there is nothing to be gained from reproducing them here.

Microphones and loudspeakers

These transducers are well known as they appear in so many items of electronic equipment: tape recorders, radio, TV, record players, and public address systems, to name only a few. This is a good opportunity to describe the use of transducers in a loudspeaker system.

In the last chapter we came across a quantity called 'impedance'; it's a combination of resistance and reactance, and loudspeakers are sold not only in terms of their size, bandwidth and power handling capacity, but

Table 12.1 Some examples of transducers

Transducer	*Input energy*	*Output energy*
Loudspeaker	Electrical	Acoustical
Microphone	Acoustical	Electrical
Light bulb	Electrical	Light
Neon lamp	Electrical	Light
LED	Electrical	Light
Photo diode	Light	Electrical
Transformer	Electrical	Electrical
Aerial	Electromagnetic	Electrical
Antenna	Electrical	Electromagnetic
Tape Heads		
Record	Electrical	Magnetic
Playback	Magnetic	Electrical
DC motor	Electrical	Mechanical
Generator	Mechanical	Electrical
7-segment display	Electrical	Light
Relay		
Coil	Electrical	Mechanical
Contacts	Electrical	Electrical
Solar cell	Light	Electrical
Thermocouple	Heat	Electrical
Telephone		
Microphone	Mechanical	Electrical
Receiver	Electrical	Mechanical
Logic gate	Logic level	Logic level
TV camera	Light	Electrical
Cathode ray tube (CRT)	Electrical	Light

also their impedance. Typical values include 3 Ω, 8 Ω, 15 Ω and 80 Ω and it is very important to obtain the correct impedance value so that it 'matches' the output of the amplifier it's connected to. This was mentioned in Chapter 4.

Microphones

A good microphone must be able to respond to all the audio frequencies, i.e. 20 Hz to 20 kHz. There are various different types.

(a) Moving coil or dynamic microphone

This is one of the most popular types of microphone as it gives good quality reproduction, is fairly robust and has omnidirectional properties, that is to say, it can pick up sounds in all directions. As the name 'moving coil' implies, a coil is made to move inside a magnetic field as someone speaks into the microphone. This type is therefore the opposite in operation to a loudspeaker. The impedance ranges from as little as 50 Ω to around 600 Ω. This is considered to be a low impedance, so these microphones must be used with an amplifier which has a low impedance input. If this is not available, the microphone must be matched either by using a specialised transformer or an appropriate pre-amplifier.

(b) Carbon microphones

These are often used in telephones. As long as the communication is intelligible, hi-fi reproduction is not required. Although the bandwidth may be restricted to about 30 Hz to 3 kHz, the quality is high enough, not only to understand what is being said, but also to be able to recognise the speaker's voice in most cases.

(c) Crystal microphones

These are in common use in reel-to-reel and cassette recorders. The quality of reproduction is generally regarded as being sufficient for home recordings. Crystal microphones usually have a very high impedance of between 1 MΩ and 5 MΩ.

(d) Capacitor microphones

These are used where the quality of reproduction is most important. Common applications include radio and TV stations and PA systems used at concerts.

Some common symbols are shown in Figure 12.2.

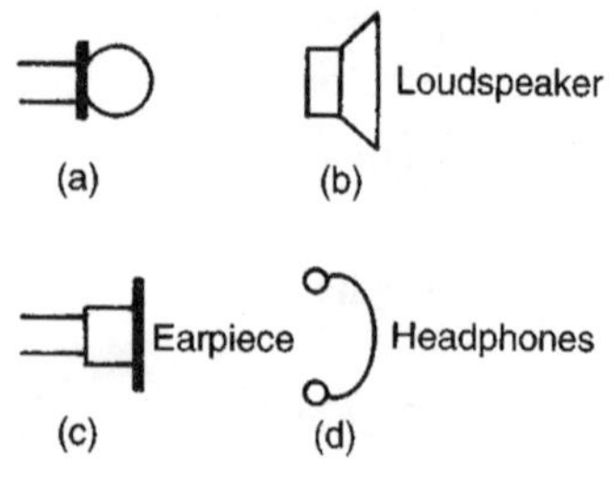

Figure 12.2 Symbols for (a) a microphone, (b) a loudspeaker, (c) an earpiece and (d) headphones

Loudspeakers

Most common loudspeakers are low impedance types. In the days of valve amplifiers, an output transformer would generally be matched to a 3 Ω loudspeaker. Nowadays, where transistor amplifiers dominate, common impedances are 8 Ω, 15 Ω and up to 80 Ω or more.

Very large loudspeakers may be used to emphasise low (bass) frequencies; these are called 'woofers'. Very small loudspeakers, called 'tweeters', are used to emphasise the high (treble) frequencies.

(a) Headphones

Those used for stereo systems and portable devices, like personal tape and CD players, tend to be low impedance (around 8 Ω) moving coil types. Lower quality headsets may have impedances of around 1 kΩ.

(b) Earpieces

These are less common than they used to be, although they are still used in hearing aids. There are two main types, low impedance (about 8 Ω)

and high impedance (1 MΩ to 5 MΩ). The high impedance types must be used with 'crystal sets' such as the one described in Chapter 7.

Public address systems

It has already been stated that transducers must be 'matched' to the input or output of the system to which they are connected. In a public address (PA) system, the microphone must be matched to the input of the amplifier and the loudspeaker must be matched to its output.

A typical PA system is shown in Figure 12.3.

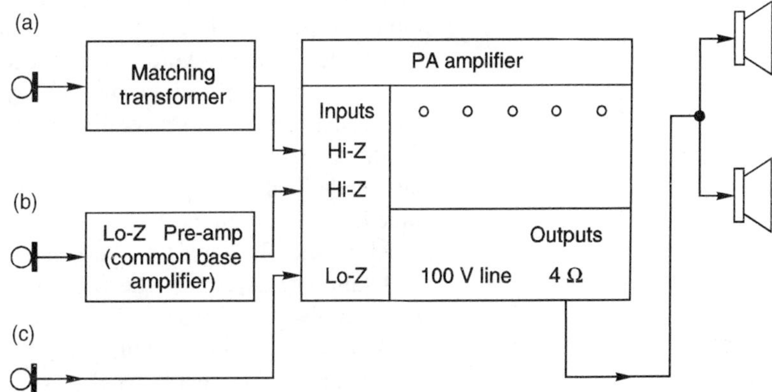

Figure 12.3 Transducers in a PA system. Three microphones are being used, all low impedance. The top one (a) uses a matching transformer, (b) uses a pre-amp and (c) goes directly to the low impedance input of the amplifier. Two 8 Ω loudspeakers in parallel are connected to the 4 Ω output

There are three microphones, all of them low impedance types. They convert sound energy into electrical energy. The amplifier has three inputs, two of them are high impedance and one is low impedance, so the low impedance microphone may be plugged into the amplifier directly. One of the microphones is 'matched' to the amplifier using a special transformer, the other is matched by using a transistor pre-amplifier in the common base configuration.

The amplifier has a 4 Ω output which will therefore be perfectly matched to the two 8 Ω loudspeakers connected in parallel, as shown in the diagram. A '100 V' line output is also shown; this is a special output which is designed to feed loudspeakers which may be some distance from the amplifier. Each loudspeaker connected to such an output must have its own '100 V line' transformer, which is generally connected inside the loudspeaker enclosure. Not only does this allow longer runs to be connected, but also removes the necessity for impedance matching.

You may have noticed that the impedance of some loudspeakers is very similar to that of some microphones. Moving coil microphones are available with low impedances such as 50 or 60 Ω and many loudspeakers have a similar value.

A loudspeaker works on the principle of a current-carrying conductor in a magnetic field experiencing a force; moving coil microphones work on a complementary principle – that of producing an emf by the movement of a conductor in a magnetic field. Since both instruments work on a similar principle, could one be used as the other?

The answer to the question is yes. A loudspeaker can be made to work as a microphone, and this fact is used to great advantage in many, simple intercom systems. Switching enables the loudspeakers at each end of the system to function in the normal way or as microphones. If a crystal microphone (impedance 1 MΩ or more) is connected to an oscilloscope, it is possible to see the waveforms produced as you speak. Connecting a loudspeaker in the same way has hardly any effect, because most loudspeakers have low impedances (generally just a few ohms, as we have seen). This is where matching – using an appropriate transducer – is necessary. A purpose-built matching transformer should be used for connecting a low impedance microphone to a high impedance amplifier input, but many transformers will work well enough to demonstrate the principle.

Obtain a transformer whose secondary is designed to produce around 6 V from the 240 V mains. *Remember at all times*, that the transformer is being used for a different purpose here and should in *no way* be connected to the mains. Connect a loudspeaker (almost any will do) to the 6 V secondary and connect what should be the mains input to the oscilloscope. If it is correctly adjusted, the oscilloscope will produce audio waveforms as you tap or speak into the loudspeaker.

If a small audio amplifier is available, it's a simple matter to design and set up an intercom system using two loudspeakers, matching transformers and the necessary switching. Much can be learned from doing this. The basic idea is shown in Figure 12.4.

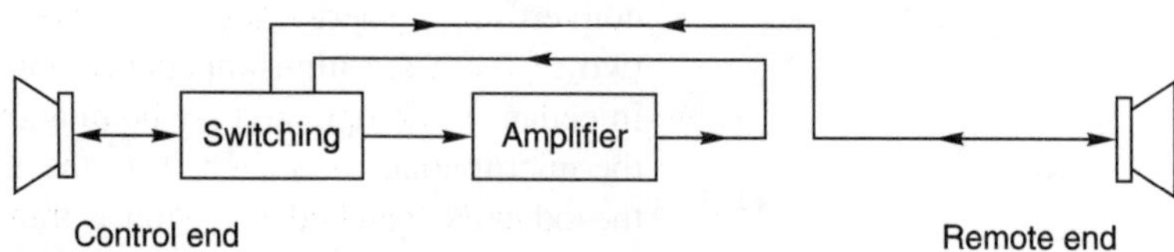

Figure 12.4 Basic circuitry for a two-way intercom system

7-segment displays

These displays, which are designed to produce the numbers 0 to 9, simply consist of an array of seven light-emitting diodes, eight if you count the decimal point. They are usually available in a dual-in-line, plastic (DIP) package. For compactness, one side of all the diodes is 'commoned up', that is to say, all the anodes are connected together, or all the cathodes are connected together. Hence, these devices are called either 'common anode' or 'common cathode' types. Figure 12.5(a) shows the display with each

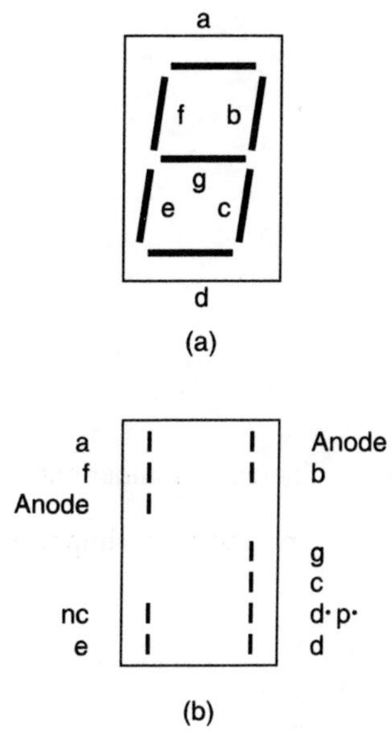

Figure 12.5 (a) The display with each segment denoted by a small letter of the alphabet. (b) The pin-out for a common anode type

segment denoted by a (small) letter of the alphabet, while Figure 12.5(b) shows a typical pin-out.

A theoretical diagram of a common anode 7-segment display is shown in Figure 12.6(a) while the common cathode type is shown in Figure 12.6(b).

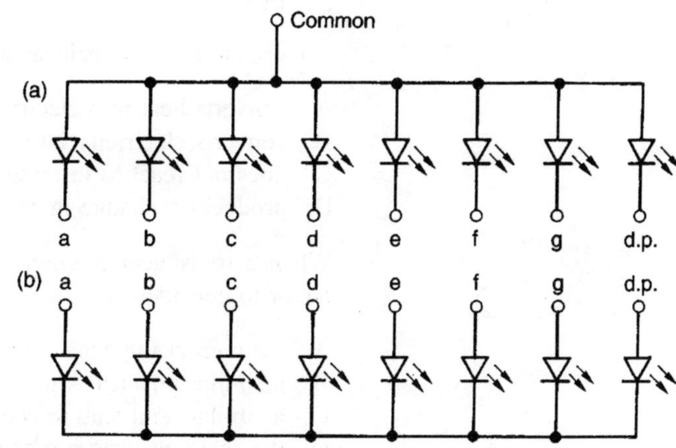

Figure 12.6 (a) The common anode connections of a 7-segment display. (b) The common cathode connections of a 7-segment display

A quick look at Table 12.1 (if this is necessary) will provide details of the input and output energies of these display transducers. Where they are needed at this level, all the other transducers mentioned in the table are described elsewhere in this book. If any difficulty arises, the index will show the way.

Multiple choice questions

1 One example of an electroacoustic transducer is

A a transformer
B a transistor
C a loud hailer
D a loudspeaker

2 A light dependent resistor (LDR)

A produces a current when light shines on it
B varies its resistance when light shines on it
C produces light when a current flows through it
D only reacts to infrared radiation

3 A seven segment display may be regarded as

A an output transducer
B an input transducer
C a light sensitive transducer
D a temperature sensitive transducer

4 A typical loudspeaker impedance value is

A $0.15\,\Omega$
B $15\,\Omega$
C $15\,k\Omega$
D $15\,M\Omega$

5 A thermistor is a transducer which

A converts heat into electricity
B converts electricity into heat
C does not react to infrared radiation
D produces a change in resistance with a change in temperature

6 When a transducer is connected into a system, the most important factor to consider is

A the correct impedance matching
B a similar physical size
C a similar temperature coefficient
D the same electrostatic compatibility

7 One type of seven segment display may be described as

A common cathode
B common base
C common emitter
D common collector

8 Good microphones must be able to respond to frequencies in the range

A 2 Hz to 20 Hz
B 20 Hz to 2 kHz
C 20 Hz to 20 kHz
D 20 kHz to 20 MHz

9 A loudspeaker may be used to display a 'picture' of the human voice on an oscilloscope provided that

A you shout loud enough
B the loudspeaker has a very low impedance
C the loudspeaker has a cone diameter of at least 12″
D a suitable matching transformer is used

10 Examples of transducers which convert light energy into electrical energy are

A solar cells, photo diodes and TV cameras
B solar cells, photo diodes and LEDs
C solar cells, thermocouples and thermistors
D video cameras, cathode ray tubes and photo diodes

11 A transducer which can convert sound vibrations into electrical energy is a

A homophone
B microphone
C megaphone
D xylophone

12 The transducer usually used at the output of an audio amplifier is

A a piezo-electric device
B an audible warning device (AWD)
C a microphone
D a loudspeaker

Part Three
Electronics Applications

13 Mathematics for electronics

Basic arithmetic

It is not unfair to assume that you are already familiar with the basic operations of addition, subtraction, multiplication and division; however, as a starting point some simple examples with the numbers 492, 385 and 38 are included.

Addition:

$$\begin{array}{r} 492 + \\ 385 \\ \hline 877 \\ \hline \end{array}$$

Subtraction:

$$\begin{array}{r} 492 - \\ 385 \\ \hline 107 \\ \hline \end{array}$$

Multiplication:

$$\begin{array}{r} 492 \\ 385 \\ \hline 147\,600 \\ 39\,360 \\ 2\,460 \\ \hline 189\,420 \\ \hline \end{array}$$

Division:

$$\begin{array}{r} 12.9473 \\ 38\sqrt{492.0000} \\ 38 \\ \hline 112 \\ 76 \\ \hline 360 \\ 342 \\ \hline 180 \\ 152 \\ \hline 280 \\ 266 \\ \hline 140 \\ 114 \\ \hline \text{etc.} \end{array}$$

Percentages

Per cent means per hundred:

15% means $\frac{15}{100} = \frac{3}{20}$ as a fraction.

More difficult percentages may be turned into fractions by multiplying the fractional part with a number that will eliminate the fraction. Here's an example:

$$83\tfrac{1}{3}\% = \frac{83\frac{1}{3}}{100} = \frac{83\frac{1}{3} \times 3}{100 \times 3}$$

$$= \frac{250}{300} = \frac{5}{6} = 0.833\ldots$$

Similarly, a decimal percentage can be converted as follows:

$$9.6\% = \frac{9.6}{100} = \frac{96}{1000} = \frac{12}{125} = 0.096$$

And the result shows that it may be obtained by simply shifting the decimal point two places to the left.

Changing fractions or decimals to a percentage

To convert a fraction or a decimal into a percentage, simply multiply it by 100:

$$\frac{3}{5} = \left[\frac{3}{5} \times 100\right]\% = \frac{300}{5} = 60\%$$

Ratio

To find a ratio, let's say between 375 and 500, write the numbers down as a quotient and look for a common denominator. In this example, it is obvious that both numbers can be divided by 5. Successive divisions will leave the fraction:

$$\frac{375}{500} = \frac{75}{100} = \frac{15}{20} = \frac{3}{4} \quad \text{i.e. a ratio of 3:4}$$

Average

Simple averages may be calculated by adding together all the given numbers and then dividing by the number of numbers there are.

Calculate the average value of the following numbers:

3, 45, 34, 6, 33, 21, 8, 4, 56, 21

In this example there are ten numbers, so the average value can be calculated by adding all the numbers together and dividing by ten:

The sum is: 231, so the average is this divided by 10,

i.e. average = 23.1

Tolerance

This subject is covered in Chapter 14 where the tolerance of resistors is discussed. Here are a few more examples.

In the E12 series, the tolerance is given as ±10%. What are the maximum and minimum values of a 39 Ω resistor?

The maximum value will be 39 Ω + (10% of 39)

i.e. 39 Ω + 3.9 Ω = 42.9 Ω

The minimum value will be 39 Ω − (10% of 39 Ω)

i.e. 39 Ω − 3.9 Ω = 35.1 Ω

The range, therefore, is 35.1 Ω to 42.9 Ω

Significant figures

Basically, this is just a method of rounding up or down. For example, what is 35 892 to two significant figures?

Answer

35 892 to 2 s.f. is 36 000

to 3 s.f. is 35 900

To take another example, what is 4279 to 1, 2 and 3 significant figures?

Answer

4279 to 1 s.f. is 4000

to 2 s.f. is 4300

to 3 s.f. is 4280

Decimal places

This procedure is basically the same except that it involves rounding up or down a number of decimal places:

$\pi = 3.141592654$

Correct to:	3 d.p.	2 d.p.	1 d.p.
	3.142	3.14	3.1

Any decimal number can be quoted to so many decimal places in the same way.

Multiples of SI units

The names of the multiples and sub-multiples of the units are formed by means of the following prefixes:

Factor by which the unit is multiplied	*Prefix*	*Symbol*
$1\,000\,000\,000\,000\,000\,000 = 10^{18}$	exa	E
$1\,000\,000\,000\,000\,000 = 10^{15}$	peta	P
$1\,000\,000\,000\,000 = 10^{12}$	tera	T
$1\,000\,000\,000 = 10^{9}$	giga	G
$1\,000\,000 = 10^{6}$	mega	M
$1\,000 = 10^{3}$	kilo	k
$100 = 10^{2}$	hecto	h
$10 = 10^{1}$	deca	da
$0.1 = 10^{-1}$	deci	d
$0.01 = 10^{-2}$	centi	c
$0.001 = 10^{-3}$	milli	m
$0.000\,001 = 10^{-6}$	micro	μ
$0.000\,000\,0001 = 10^{-9}$	nano	n
$0.000\,000\,000\,001 = 10^{-12}$	pico	p
$0.000\,000\,000\,000\,001 = 10^{-15}$	femto	f
$0.000\,000\,000\,000\,000\,001 = 10^{-18}$	atto	a

Examples

Current: 0.01 A = 10^{-2} A = 10 mA; 0.1 A = 10^{-1} A = 100 mA

Capacitance: 1 μF = 10^{-6}; 1 nF = 10^{-9} F;

1 pF = 10^{-12} F; 0.1 μF = 100 nF

Inductance: 1 mH = 10^{-3} H; 1 μH = 10^{-6} H;

10 mH = 10^{-2} H or 0.01 H

Frequency: 1000 c.p.s. or c/s = 1 kHZ or 10^3 Hz;

10^6 Hz = 1 MHz; 1000 MHz = 1 GHz

Resistance: 1000 Ω = 1 kΩ; 10^5 Ω = 100 kΩ; 10^6 Ω = 1 MΩ

2200 Ω = 2.2 kΩ or 2K2 if marked on the resistor

Standard form

This is a standard method of writing numbers down and is especially useful if they are very small or very large. A number in standard form consists of a positive number equal to 1 or more, but less than 10. An example may be helpful:

The number 2385 would be: 2.385×10^3

The number 70 000 would be: 7×10^4

The number 74.29×10^5 is not in standard form because the first two digits (74) produce a number greater than 9. Similarly, 0.26×10^6 is not in standard form because 0.26 is less than 1.

Changing the subject of a formula

This technique is also called 'transposition'. Take the formula for calculating the value of a capacitor from its physical dimensions:

$$C = \frac{\epsilon_o \epsilon_r A}{d}$$

Now if we want ϵ_o to be the subject of the formula, one way of doing it is as follows: first make sure that the quantity required is above the line; in this case it is, so put the whole lot on the left hand side, as follows:

$$\frac{\epsilon_o \epsilon_r A}{d} = C$$

From then on it's easy; leave ϵ_o where it is and then move anything on the top of the left hand side to the bottom on the right hand side:

$$\frac{\epsilon_o}{d} = \frac{C}{\epsilon_r A}$$

Now move everything on the bottom of the left hand side to the top on the right hand side:

$$\epsilon_o = \frac{Cd}{\epsilon_r A}$$

And that's all there is to it.

Questions

1 Without the use of a calculator, take the numbers 1894 and 18 and:

(a) add them together
(b) subtract the smaller from the larger
(c) multiply them together
(d) divide the smaller into the larger.

2 Calculate 10%, 15%, 30% and 75% of the following numbers:

(a) 50
(b) 88

3 (a) Write down the following numbers to 2 significant figures:

(i) 3288 (ii) 3945 (iii) 1672
(iv) 630 (v) 29.5 (vi) 0.0881

(b) Correct the following numbers to 2 decimal places:

(i) 3.14159 (ii) 0.815 (iii) 15.673
(iv) 8.999 (v) 0.0763 (vi) 144.2

4 Put the following numbers into standard form:

(i) 6492 (ii) 1 100 000 (iii) 560 000
(iv) 22 (v) 0.055 (vi) 0.0075
(vii) 27 000 000 (viii) 59637 (ix) 1.33
(x) 0.00009

5 (a) The following resistors have a tolerance of 10%. State the RANGE of values each resistor may have.

(i) 1 kΩ (ii) 27 kΩ (iii) 56 kΩ
(iv) 220 kΩ (v) 3.3 MΩ

(b) The following resistors have a tolerance of 5%. State the RANGE of values each resistor may have.

(i) 910 kΩ (ii) 75 Ω (iii) 22 kΩ
(iv) 2.2 MΩ (v) 15 kΩ

6 Transpose the following formulae making the value given after each equation the subject of the formula:

Example: $C = \dfrac{\epsilon_o \epsilon_r A}{d}; A$

becomes: $A = \dfrac{Cd}{\epsilon_o \epsilon_r}$

(i) $I = \dfrac{V}{R}; R$

(ii) $C = \dfrac{Q}{R}; Q$

(iii) $P = IV; I$

(iv) $P = I^2 R; I$

(v) $e = mc^2; c$

14 The resistor colour code

Examples

Most carbon resistors in common use have up to four coloured bands on them. The first three bands give the value of the resistor in ohms (Ω) and the fourth band gives the value of something known as 'tolerance'. To read the value, you take the first two coloured bands and note what value the colours represent, then the third band gives the 'multiplier', i.e. the number of zeros which follow (more technically, the power of ten to which the first two figures are multiplied). A few examples should be helpful, but first here is a list of the colours used with their values:

Black	– 0	Green	– 5
Brown	– 1	Blue	– 6
Red	– 2	Violet	– 7
Orange	– 3	Grey	– 8
Yellow	– 4	White	– 9

So, if a resistor has the code sequence red, red, orange, then this corresponds to 22 (for the two reds) followed by orange (the multiplier). Orange is 3, which means the 22 is followed with three zeros, or, as indicated above, the multiplier as a power of ten, which means the 22 is multiplied by 10^3 (one thousand). In either case a value of 22 000 Ω or 22 kΩ is obtained. The names of the multiples and sub-multiples of units were given in the last chapter and now is a good opportunity to start using them.

Further examples

The sequence blue, grey, red produces 68 followed by two zeros, i.e. 6800 Ω or 6.8 kΩ. Or, it is 68 multiplied by 10^2 which obviously gives the same answer – $68 \times 100 = 6800$ Ω.

Grey, red, orange becomes 82 000 Ω or 82 kΩ, and green, white, black is 59 Ω. In this example, we have 59 followed by the number of zeros indicated by the colour black. Black is zero, so 59 is followed by no zeros – hence, 59 Ω. Using the other method, we have 59 multiplied by 10^0 which is 1, and 59×1 is 59. (Any number to the power zero is equal to one.)

Not all resistors are marked with a colour code; wirewound resistors, for example, which are designed to carry much more current than carbon types, usually have the value printed on them. Some manufacturers now use the BS 1852 resistor code and some examples of this are given below:

0.47 Ω	would be marked	R47
1 Ω	would be marked	1R0
4.7 Ω	would be marked	4R7
47 Ω	would be marked	47R
100 Ω	would be marked	100R
1k Ω	would be marked	1K0
10k Ω	would be marked	10K
10M Ω	would be marked	10M

Tolerance

Generally, the lowest common value of carbon resistor is about 1 Ω and the highest 27 MΩ. It is possible to buy smaller and larger values, of course, but these are usually for specialised applications and would rarely be needed. This means that when a technician wishes to buy various value resistors, even holding a stock of just one of each type would seem to indicate that 27 million resistors would be held in stock! This is, of course, ridiculous but fortunately quite unnecessary. This is where the tolerance of the resistor comes in. A resistor sold with a value of, say, 100 Ω has a *nominal* value of 100 Ω. In actual fact, if it has a tolerance of ±10% the value could be anywhere between 90 Ω and 110 Ω.

Given that for most practical purposes, the precise value of a resistor is unimportant, there is no point in making 91 Ω, 92 Ω and 93 Ω etc., because a resistor of 100 Ω would be near enough. So, the next value made above 100 Ω is 120 Ω; this is because 100 + 10% overlaps with 120 − 10% and the larger the number gets, the bigger the gap between succeeding values. It turns out that in the series of resistors with ±10% tolerance, only twelve, basic numerical values exist; all the other values are simply multiples of them. Because there are twelve, this is called the E12 series and the values are as follows:

E12 (±10%) 10 12 15 18 22 27 33 39 47 56 68 82

Resistors with a tolerance of ±5% occupy the E24 series and their values are given below:

E24 (±5%) 10 11 12 13 15 16 18 20 22 24 27 30
33 36 39 43 47 51 56 62 68 75 82 91

Tolerance values

The fourth coloured band on a resistor indicates the tolerance as follows:

Colours		*Letters*
Brown	±1%	F = ±1%
Red	±2%	G = ±2%
Gold	±5%	J = ±5%
Silver	±10%	K = ±10%
No band	±20%	M = ±20%

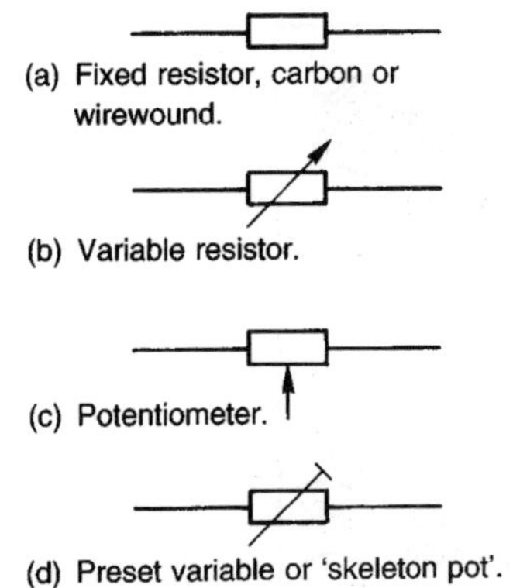

Figure 14.1 Some common resistor symbols

In practice, because there is only a limited number of resistor values available, you soon get used to the colour combinations used in the code. For example, an experienced electronics technician would be used to seeing red and violet as the first two colours in the code and if followed by a red band, the value would be 2.7 kΩ. You don't expect to see violet, red, red because 7.2 kΩ is not a common value. It certainly doesn't appear in the E12 series, or even the E24 series which gives the values of resistors with ±5% tolerance.

This means that learning the resistor colour code is not as difficult as it may seem because there's only a limited number of possibilities in everyday use. The resistors in the E12 series (and the E24 series) are called preferred values. Sometimes manufacturers quote the stability of resistors; this refers to the ability of the resistor to retain its stated value during both storage and use.

Power rating

This indicates the maximum power which may be dissipated in a component with an excessive rise in temperature. A 200 Ω resistor carrying 1 mA must dissipate 200 μW, but the same resistor carrying 100 mA must dissipate 2 W. In consequence, resistors are sold by resistance value and power rating, typical values including $\frac{1}{2}$ W, 1 W, 2 W, 3 W and 5 W. Higher values are available as wirewound resistors or metal film resistors.

There are many different types of resistor but basically they fall into two categories, carbon and wirewound. Both types are available as fixed resistors and variables, and some symbols are shown in Figure 14.1.

BS 3939 symbols and IC pin-outs

There are many conventions for drawing electronic circuits and these tend to vary largely with age or country of origin. Many circuit symbols in general use are described in this chapter but no one is suggesting that they should be learned off by heart. Familiarity comes with frequent use and you should have come across many of them before already. In Figure 14.2 are some of the most common circuit symbols and conventions.

It is not unusual to see old electronic symbols still in use; at the time of writing, for example, *Television* magazine continues to use the 'zig-zag' symbol for a resistor, rather than the British Standard rectangular box. On the other hand, the American Mil. Spec. logic gate symbols are much more popular than the British Standard ones. Standard logic symbols are given in Chapter 9 (page 140).

All dual-in-line (DIL) IC packages have a standard convention for pin numbering. They start with pin 1 (usually indicated by a dot) in the top left hand corner, continue down, across and then up to the top right hand corner. A typical 8-pin DIL package is shown in Figure 14.3. Common examples of the use of this package include the 555 timer and the 741 op-amp.

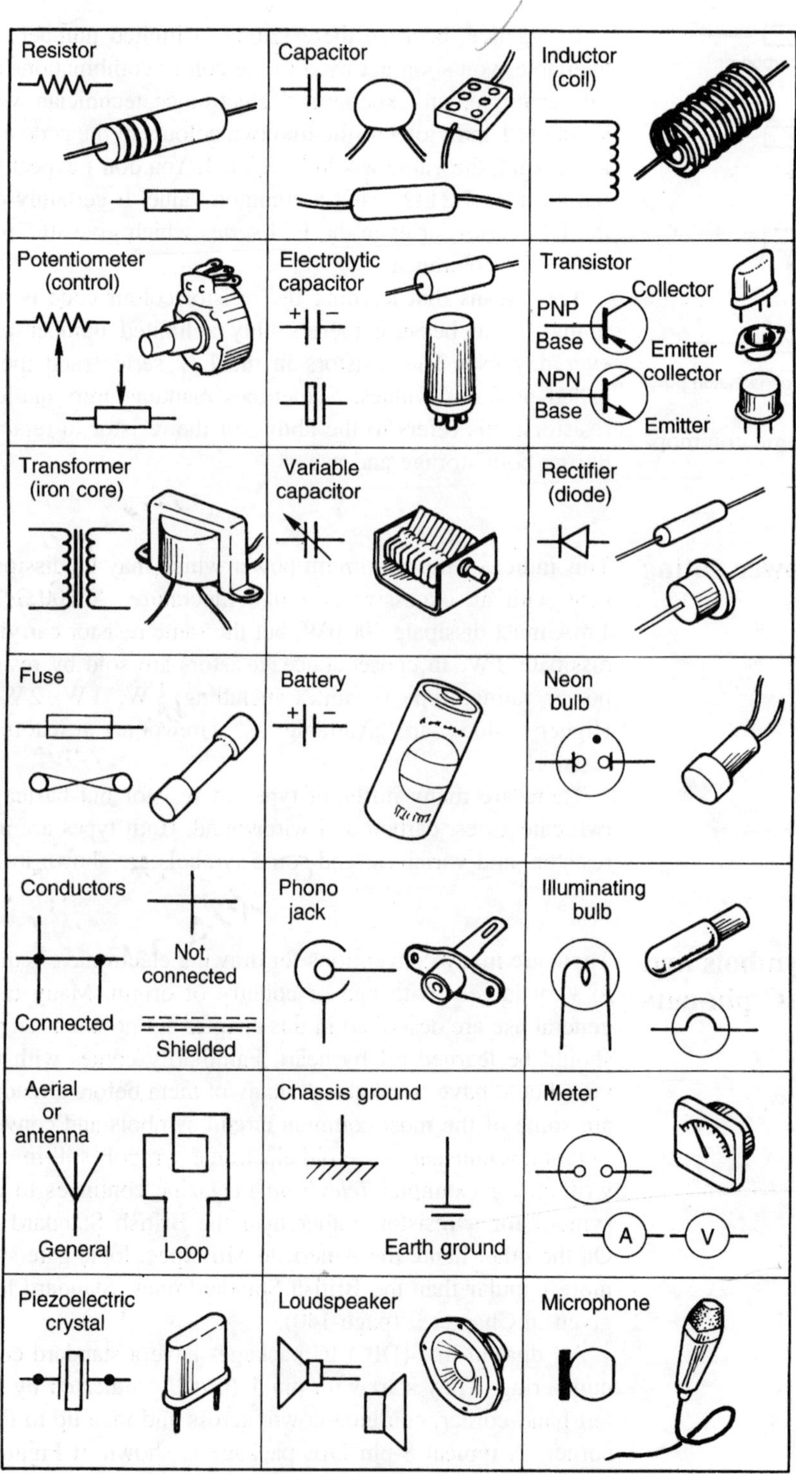

Figure 14.2 Some of the most common circuit symbols and conventions

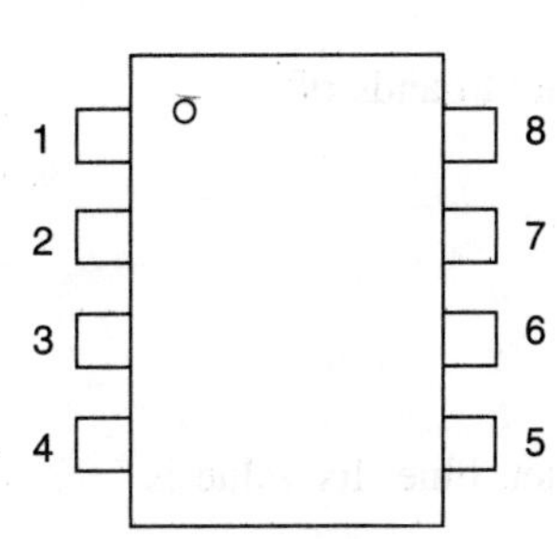

Figure 14.3 The standard 8-pin DIL IC package

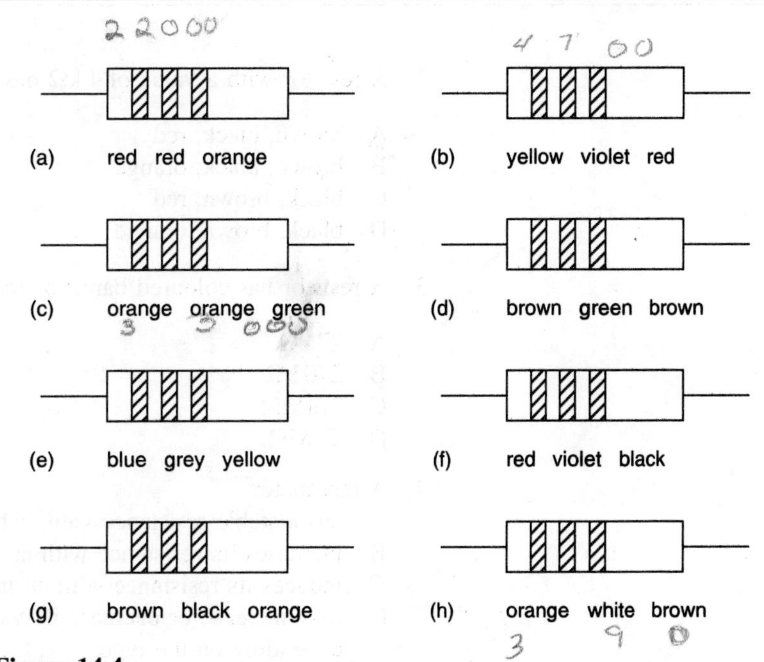

Figure 14.4

Questions

1 Write down the values of the resistors illustrated in Figure 14.4.

2 Write down the colour codes of the resistors listed below:

(a) 18 kΩ	(b) 82 Ω	(c) 680 Ω
(d) 3.3 kΩ	(e) 910 kΩ	(f) 5.6 MΩ
(g) 460 Ω	(h) 10 Ω	(i) 56 Ω
(j) 270 Ω	(k) 22 kΩ	(l) 470 kΩ

3 Choosing resistors from the E12 series, which standard values are closest to those given below?

(a) 90 Ω (b) 21 Ω (c) 500 Ω

4 A resistor carries a current of 200 mA when a potential difference of 5.4 V exists across it. What is the colour code of the resistor?

5 A resistor with the colour code: red, red, red, has a potential difference of 15 V across it. How much current flows through it?

Multiple choice questions

1 A resistor has four coloured bands as follows: red, red, orange, gold. Its tolerance is

A ±2%
B ±5%
C ±10%
D ±20%

2 A resistor with a value of 1 kΩ has coloured bands of

A brown, black, red
B brown, black, orange
C black, brown, red
D black, brown, orange

3 A resistor has coloured bands of red, violet, blue. Its value is

A 27 kΩ
B 270 kΩ
C 2.7 MΩ
D 27 MΩ

4 A thermistor
A has a stable resistance value when heated
B increases its resistance with an increase in temperature
C reduces its resistance with an increase in temperature
D may increase or decrease its value with a change in temperature, depending on the type

5 When a piece of copper wire is heated its resistance

A increases
B decreases
C stays the same
D becomes infinite

15 Graphs and graphical representation; waveform measurement using a CRO

Graphical representation

Graphs are a very useful and convenient method of presenting data and obtaining further information from it. Generally, two axes are used, the x-axis which is drawn horizontally and the y-axis, which is drawn vertically. The quantity represented by the x-axis is known as the independent variable; for example, if an experiment is designed to show how the resistance of a conductor varies with temperature, then it's the temperature which is the independent variable. The resistance varies with the temperature; it is not the temperature which varies according to a change in resistance.

Graphs should be drawn as large as possible using a (simple) scale which makes the best use of the paper. A title should be given which describes the relationship between the quantities being represented. Each axis should be clearly labelled with the quantity, its symbol and its units, and the scale used. The co-ordinates should be plotted using a cross like a plus (+) sign; diagonal crosses like multiplication signs or dots (with or without circles around them) should be avoided.

As an example, let's conduct experiment designed to show how the potential difference (voltage) across a conductor varies with the current flowing through it. A suitable circuit is shown in Figure 1.12 (Chapter 1) which may be used to verify Ohm's law.

The data are first recorded in a table such as the one shown in Table 15.1. This gives typical values in a suitable layout.

Table 15.1 Table of V and I values

Voltage (V) (volts)	*Current (I)* (amps)
1	0.10
2	0.20
3	0.29
4	0.38
5	0.50
6	0.61
.	.

A graph using the data in Table 15.1, and observing the principles previously described, is shown in Figure 15.1.

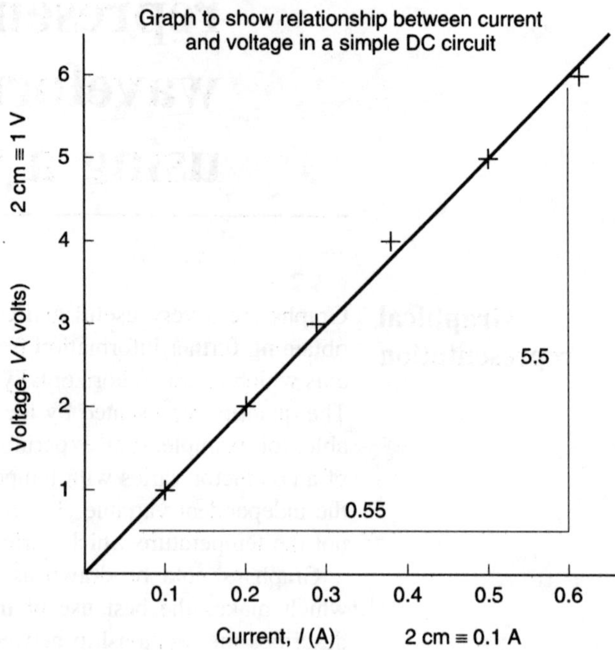

Figure 15.1 Graph drawn from the values given in Table 15.1

The voltage is plotted along the y-axis while the x-axis represents the current. The correct labels, including a suitable scale, are shown. The title is given as: 'Graph to show relationship between current and voltage in a simple DC circuit' Once the co-ordinates have been plotted, the *best* straight line (going through the origin where appropriate) is drawn in. The origin is generally at the zero value of each of the quantities being represented; in Figure 15.1 it is where the x and y values cross over in the bottom left hand corner.

The gradient of the graph is obtained by drawing in the largest triangle possible; from this, the y value is divided by the x value. Since these represent voltage and current respectively, voltage is being divided by current and the gradient is then the value of resistance in the circuit. You should see that this agrees with the formula $R = V/I$, an expression of Ohm's law (assuming a metallic conductor at constant temperature).

When such V/I graphs are usually presented, they are drawn the opposite way round, that is to say, with current being represented on the (vertical) y-axis and voltage on the x-axis. In such so-called I/V diagrams, the gradient then gives $1/R$ – the reciprocal of resistance, rather than R itself.

If values of V and I are obtained for both directions of current as shown in Table 15.2, then any graph drawn up from these data will have an origin in the centre as shown simply in Figure 15.2. You should draw up a suitable graph from the data in Table 15.2 and hence confirm that the gradient is 1/100 (using the I/V diagram convention described above) and that the value of resistance in the circuit used to obtain the data is 100 Ω.

Table 15.2 Table of V and I values, including those obtained when the current in the circuit is reversed

Voltage (V) (volts)	*Current (I)* (mA)
+5.0	50.0
4.5	44.2
4.0	39.5
3.5	34.0
3.0	29.0
2.5	24.0
2.0	19.5
1.5	14.5
1.0	11.0
0.5	5.2
0	0
−0.5	5.4
−1.0	11.0
−1.5	14.5
−2.0	21.0
−2.5	25.0
−3.0	29.5
−3.5	35.0
−4.0	39.5
−4.5	45.0
−5.0	50.0

The graphs described so far are linear ones, that is to say, they are straight line graphs. One quantity is directly proportional to the other; in our first example, if the voltage across a resistor is doubled, then the current flowing through it is also doubled. Some quantities in electronics are inversely proportional to each other, the variation of NTC thermistor resistance with temperature being a good example.

A typical thermistor might have a resistance of 15 kΩ at 0°C and measure only 320 Ω at 100°C. As the temperature increases, the resistance reduces and it is not a linear relationship. If a graph is drawn from the data supplied in Table 15.3 a curve will be produced and you are encouraged to do this.

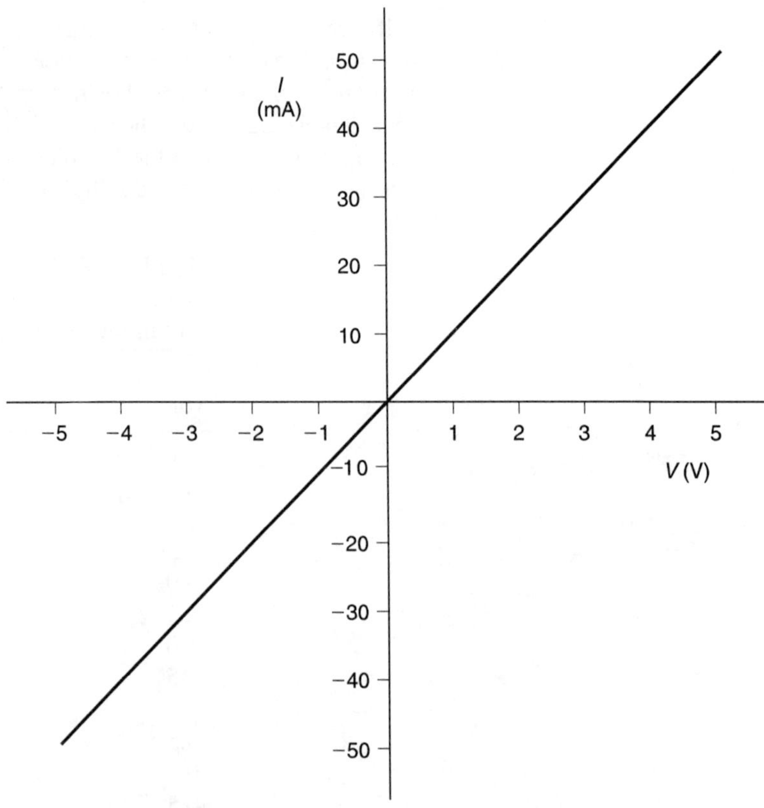

Figure 15.2 Graph drawn from V and I values given in Table 15.2

Table 15.3 Table of temperature and corresponding resistance values for a typical thermistor. Values at 0°C, 25°C and 100°C courtesy of Maplin Electronics, other values estimated for purposes of the exercise

Temperature t (°C)	*Resistance* R (k Ω)
0.0	15.28
10.0	7.00
25.0	4.70
40.0	3.50
60.0	1.75
80.0	1.00
90.0	1.20
100.0	0.32

Exponential relationships

The best example of exponential relationships in electronics is that of the charge and discharge of capacitors. All the information required is given in Chapter 3 so it is possible for you to conduct suitable experiments to obtain the necessary data from which graphs may be drawn. In addition, some data values are presented in Table 15.4.

Table 15.4 Data for capacitor charge and discharge curves. A 470 kΩ resistor and a 2.2 μF capacitor with a 9 V supply were used

CAPACITOR CHARGE		*CAPACITOR DISCHARGE*	
Time, t (seconds)	*Voltage, V* (volts)	*Time, t* (seconds)	*Voltage, V* (volts)
0	0	0	9.0
0.1	0.86	0.1	8.2
0.2	1.63	0.2	7.7
0.4	3.00	0.4	6.0
0.6	4.10	0.6	5.0
0.8	5.00	0.8	4.0
1.0	5.70	1.0	3.3
2.0	7.80	2.0	1.2
4.0	8.80	4.0	0.2
10.0	8.99	10.0	0.0004

Waveform measurement using a CRO

The oscilloscope may be regarded as an electronic device which produces a graph of voltage against time. It is therefore capable of displaying steady state DC voltages, and, more importantly, alternating voltages up to frequencies of 20 MHz in most simple oscilloscopes, 60 MHz in more expensive ones and even higher frequencies in the most specialised. The basic techniques for using a CRO and interpreting the waveforms displayed on it are, however, the same in each case.

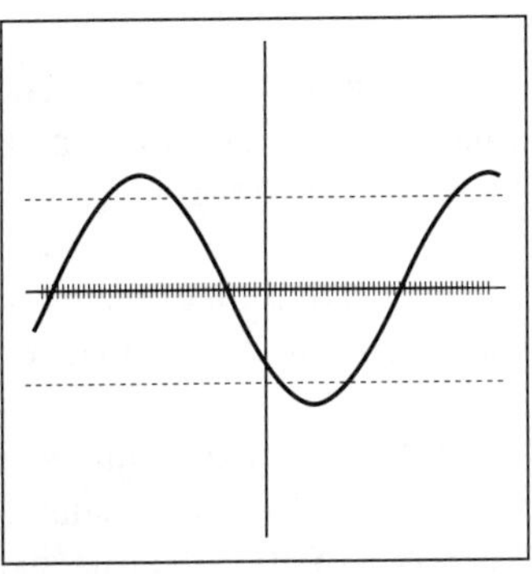

Figure 15.3 The screen of a typical CRO. A sine wave of 2.5 V amplitude and 125 Hz frequency is shown

The screen of a CRO displays its information in the same way that a normal graph does. The horizontal X-axis represents time and the vertical Y-axis represents voltage. If the timebase control is adjusted so that the spot of light on the CRO screen can be seen to move from one side to the other (the trace) then it's probably travelling at about 0.5 cm per second. As the timebase frequency is increased, the spot moves faster and faster and it is soon impossible to see the motion at all; the trace appears as a stationary, horizontal line. If a signal (say a sine wave for simplicity) is now applied to the Y input of the CRO, one or more complete cycles of the waveform may be displayed. The number of centimetres occupied by one cycle, multiplied by the timebase frequency in cm/sec, gives the period of the waveform, T (s), and the reciprocal of this ($1/T$) gives its frequency (Hz).

The amplitude, or peak value, of the waveform may be calculated by measuring the maximum height the waveform reaches above the zero line, in centimetres. This value is then multiplied by the setting of the Y amplifier control which then gives the peak voltage of the waveform. To assist in both voltage and frequency measurements, a plastic (often green) grid, called a 'graticule', divided into 1 centimetre squares and some smaller divisions, is attached to the screen of the oscilloscope tube. This is shown in Figure 15.3.

Making the measurements

Careful examination of Figure 15.3. shows that one complete cycle of the waveform occupies 8 cm. If the timebase is set to 1 ms/cm, then the period of the waveform is 8 cm × 1 ms = 8 ms. The frequency is therefore 1/8 ms which is 125 Hz. The peak value of the waveform is 2.5 cm above the 0 V line, so if the Y amplifier is set to 1 V/cm the peak voltage is 2.5 V. The period, peak and peak-to-peak values are shown in Figure 15.4.

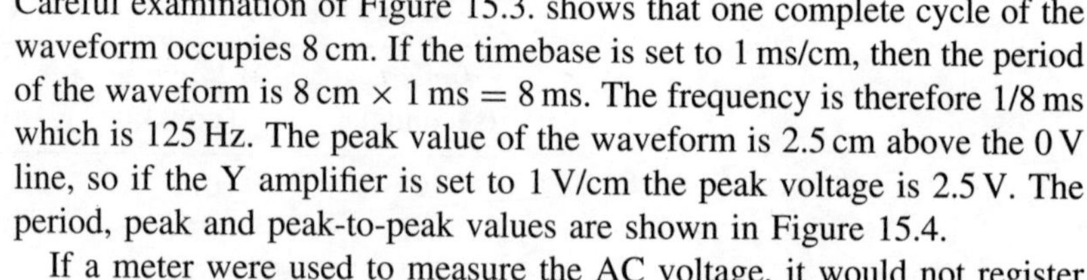

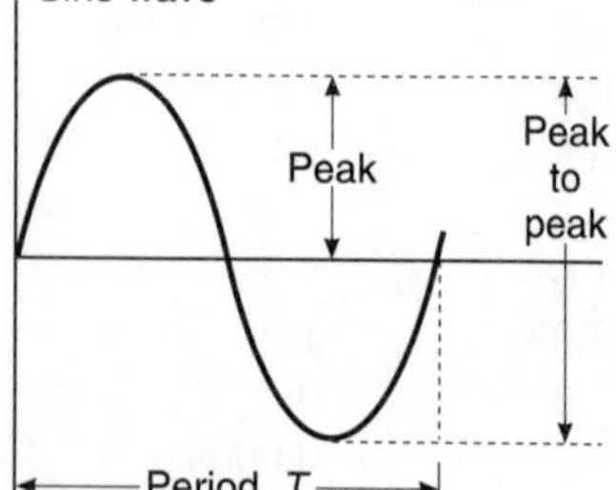

Figure 15.4 The period, peak and peak-to-peak values of a sine wave. The frequency = 1/period. Amplitude = peak

If a meter were used to measure the AC voltage, it would not register 2.5 V. The meter takes a kind of average value called the rms (root mean square) value and this is related to the peak value as shown below:

$$\text{rms} = \frac{\text{Peak}}{\sqrt{2}}$$

Just to complicate things, a value, actually called the average (or mean) value (as opposed to the rms value), also exists. This can be calculated from:

$$\text{Average} = \frac{\text{Peak}}{\pi/2}$$

Square waves

Square waves have very rapid (theoretically zero) rise and fall times. The period for which the waveform is at its maximum value is called the mark and the period for which it is at its minimum value is called the space. The addition of the mark and space times gives the period of the waveform and the reciprocal of this is, of course, the frequency. Division of the mark time by the space time gives the mark-to-space ratio. This is summarised in Figure 15.5.

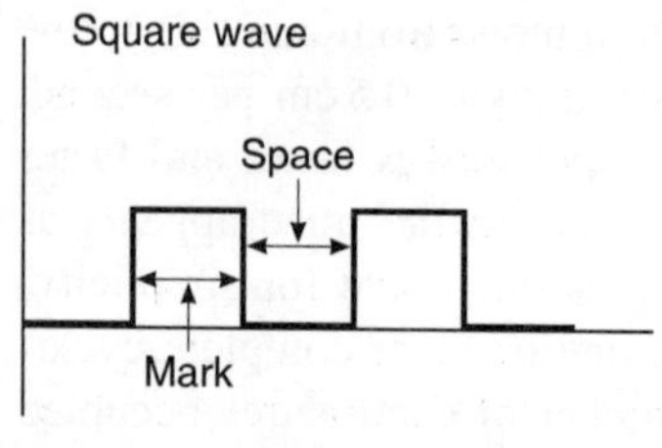

Figure 15.5 A square wave showing the mark and the space. Mark + space = period

A true square wave has mark and space intervals which are equal and therefore has a mark-to-space ratio of one. For a positive-going square wave with a mark-to-space ratio of one, (see Figure 15.6(a)) the rms and peak values are as follows:

$$\text{rms} = \frac{\text{p} - \text{p}}{\sqrt{2}} \qquad \text{Mean} = \text{Peak}/2$$

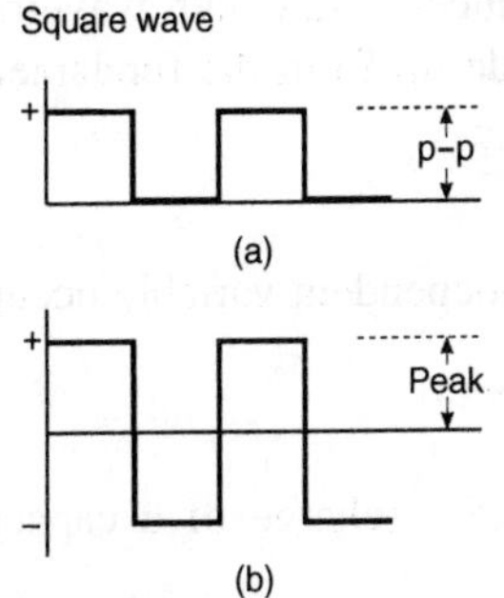

Figure 15.6 (a) A positive-going square wave with a mark-to-space ratio of one. (b) An alternating square wave

For an alternating square wave (Figure 15.6(b)), the peak, rms and mean values are all the same. The mean, or average, value over one complete cycle is, however, zero. To find out what type of wave is being considered, note the zero volt line on the CRO and switch to an AC setting.

If a waveform such as that shown in Figure 15.7 is obtained on the DC setting of the CRO, then the waveform has a DC component to it. Its average DC value is given as:

$$\text{Average DC value } = \text{Peak } - \left(\tfrac{1}{2}\text{p} - \text{p}\right)$$

This is shown in Figure 15.7.

The formula also applies to sine waves with a DC component and, to a close approximation, the ripple appearing on the output of a DC power supply.

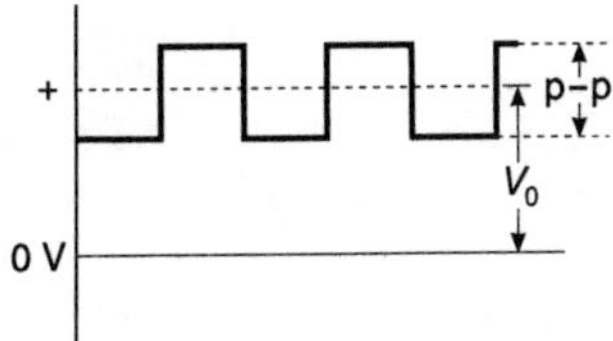

Figure 15.7 A square wave with a DC component

Time-related waveforms

Using a dual trace CRO enables two waveforms to be displayed on the screen simultaneously. This is useful for studying phase relationships and many other parameters associated with two different signals, often (but not exclusively) from the same circuit. Some time-related waveforms are shown in Figure 6.8 (Chapter 6) and Figure 8.6 (Chapter 8). Time-related waveforms may also be obtained from pins 2 and 3 of a 555 timer IC connected in astable mode and it is a very instructive and informative assignment for you to carry out.

The summation of instantaneous values

Peak, rms and mean values of a sine wave have already been described. The value of the voltage at any particular time during the cycle is called an instantaneous value; obvious examples include the peak voltage which is the maximum instantaneous value and zero voltage which is the minimum. If two sine waves are additively combined, the resultant waveform may be predicted by simply summing the respective instantaneous values of each. For example, if two sine waves in phase have peak voltages of 2 V and 3 V respectively, then the resultant waveform will have a peak voltage of 5 V.

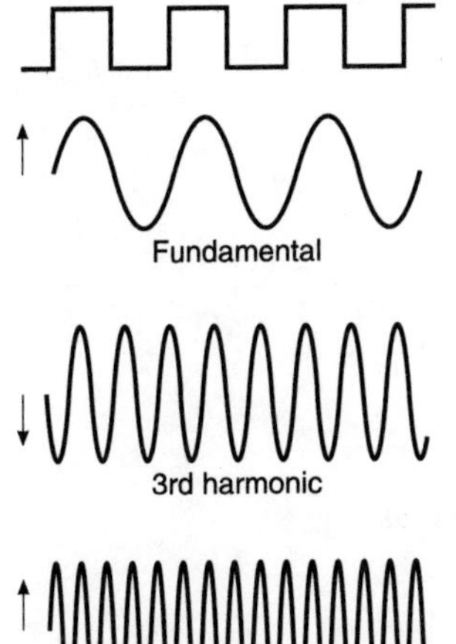

Figure 15.8 Illustrating how a square wave is made up from the fundamental and its odd harmonics

Fundamental frequency and harmonics

Any periodic waveform, regardless of its shape, may be regarded as the sum of a large number of sine waves. If a simple sine waveform is being analysed, its frequency is called the FUNDAMENTAL. Twice this frequency is called the second harmonic, three times the frequency is the third harmonic – and so on. If all the odd harmonics of a sine wave are added to the fundamental, a square wave is produced. The addition of *all* the harmonics to the fundamental produces a sawtooth waveform. Figure 15.8 shows how a square wave is made up from the fundamental and its odd harmonics.

Questions

1 On a typical graph, which axis does the independent variable occupy?

2 How is the gradient of a graph calculated?

3 What shape of graph does the charge and discharge of a capacitor produce?

4 If it takes 0.25 ms to complete one cycle, what is the frequency?

5 Values for two of: the timebase setting, distance covered by one cycle on the oscilloscope screen and frequency are given. State the third value:

(a) Timebase setting = 0.1 ms/cm
Length of cycle = 3 cm
Frequency = ?
(b) Timebase setting = 1.0 ms/cm
Length of cycle = ?
Frequency = 500 Hz
(c) Timebase setting = 5.0 ms/cm
Length of cycle = 2.5 cm
Frequency = ?
(d) Timebase setting = ?
Length of cycle = 4 cm
Frequency = 5000 Hz
(e) Timebase setting = 10 ms/cm
Length of cycle = 1.5 cm
Frequency = ?
(f) Timebase setting = ?
Length of cycle = 1 cm
Frequency = 200 Hz

Multiple choice questions

1 The amplitude of a waveform is the same as

A the peak voltage
B the peak-to-peak voltage
C the period
D the frequency

2 A waveform of frequency 5000 Hz has a period of

A 0.002 ms
B 0.02 ms

C 0.2 ms
D 2.0 ms

3 A waveform with a period of 64 μs has a frequency of

A 15.625 Hz
B 156.25 Hz
C 1.5625 kHz
D 15.625 kHz

4 A sine wave with a fundamental frequency of 1 kHz is mixed with its own odd harmonics. The resultant waveform will be most like

A a sine wave of increased amplitude
B a sine wave of reduced amplitude
C a sawtooth
D a square wave

5 One of the harmonics present in a 1 kHz square wave is

A 2 kHz
B 4 kHz
C 5 kHz
D 8 kHz

16 Electrical supplies

Energy

We have noted before (in Chapter 5) that energy is defined as the ability to do work. In order to do some work, electronic circuits must have a source of energy. This source is usually an electrical source, either in the form of batteries, or derived from the mains. There are many different forms of energy, including electrical, chemical, heat, nuclear, electromagnetic and sound energy. Light is electromagnetic radiation which is therefore a source of energy. It can be converted into electrical energy by the use of solar cells. Apart from solar cells, electricity is available from either batteries or generators.

The operation of electronic circuits

Modern, transistorised equipment usually operates on small voltages of around 12 V. Larger amplifiers may well require larger voltages. In the 'good old days' of valves, voltages of around 350 V were common, and larger amplifiers required even larger voltages. Even now, voltages of up to 25 kV are required for the tubes in televisions.

The mains supply produces alternating current (AC), nominally at 240 V, 50 Hz. All batteries produce direct current (DC). Electronic circuits need DC supplies in order that a current can be controlled, so electronic equipment must either be powered by batteries or from a DC supply derived from the mains.

Electromotive force (emf)

This is rather incorrectly named since it is not really a 'force' at all, but the energy available per coulomb of charge. A 100 V supply provides 100 joules per coulomb, for example.

The emf of a cell or battery is defined as the potential difference (p.d.) across it when an open circuit. When the battery is in circuit, the current has to flow through itself. Since the battery also has some resistance (called internal resistance), there will be a voltage dropped across it. Hence when the battery is actually being used, its p.d. will be less than its emf.

It is difficult to obtain the emf of a cell, since whenever anything is connected to it in order to measure its output, current flows and a voltage is dropped across its internal resistance. A potentiometer circuit may be used to measure an emf by the 'null' method, but generally, and especially in modern electronics, the results are of neither use nor interest. The important thing to remember is that if you buy a new battery marked, say, 1.5 V, and measure across its terminals with a digital voltmeter, then it's quite likely that you'll get more than the 1.5 V. Connect it into the kind of circuit for which it has been designed and you're more likely to get the 1.5 V advertised on the package.

Sources and types of energy supply for electronic systems

1 Mains

For all types of electronic devices, the mains will normally have to be reduced in voltage (transformed), changed to DC (rectified), and smoothed. For microprocessor systems, the smoothing must be very good, and, normally, the supply must be regulated. Power supplies are discussed more fully in Chapter 6.

2 Solar cells

These are comprised of banks of photovoltaic cells connected in series/parallel. The series connection increases the voltage available while the parallel connection increases the current available. The major application of solar cells to date is as a source of electrical power in spacecraft; for example, nearly 11 000 silicon photovoltaic cells were used on the Nimbus weather satellite producing a power output of about 0.4 W.

Recent improvements in design, especially of the lenses used in conjunction with the devices, have enabled usable powers to be developed for use in the home. It is possible, for example, to purchase battery recharging equipment which is entirely powered by the sun.

3 Batteries

There are two main types, primary and secondary. The main distinction between the two is that secondary batteries (like car batteries) are rechargeable whereas primary cells are not.

(a) Primary cells (remember that a battery consists of two or more cells). If two dissimilar metals are immersed in an acid or salt solution, called an electrolyte, an emf will be produced. This is a simple cell. In a primary cell, the chemical reaction soon ceases and the cell has to be thrown away. When the cell is producing an emf, hydrogen bubbles form and prevent the cell from operating. A depolarising agent prevents this from happening.

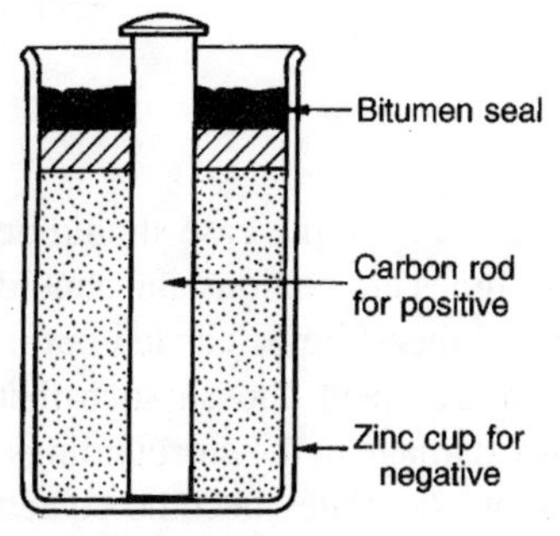

Figure 16.1 Cross-section of a typical Leclanché cell

(i) The Leclanché cell (zinc/carbon) The primary cells available at many shops are a modern version of the Leclanché cell. The anode is made from zinc, the cathode is composed of powdered manganese dioxide/carbon black wetted with electrolyte solution; a carbon rod makes electrical contact. The depolariser is manganese dioxide and the maximum emf obtainable from such cells is 1.6 V. A cross-section of a typical Leclanché cell is shown in Figure 16.1.

(ii) The alkaline/manganese cell This is an alkaline-based cell as opposed to the acid-based Leclanché cell. The anode is zinc paste while the cathode is composed of manganese dioxide and graphite with an electrolyte of potassium hydroxide. This cell usually has a steel can, the voltage available is a nominal 1.5 V and a heavy current can be supplied for long periods.

(iii) The mercury cell This has a zinc anode, a mercuric oxide cathode and the depolariser is also mercuric oxide. The electrolyte is potassium hydroxide and the whole is usually supplied in a nickel-plated steel container. The output voltage is 1.35 V to 1.4 V.

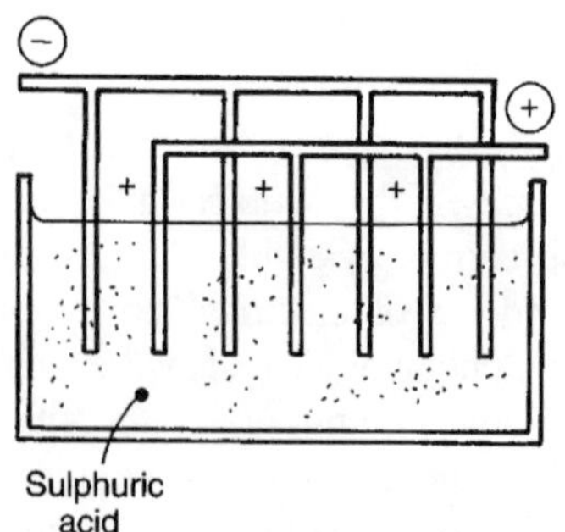

Figure 16.2 Cross-section of a lead–acid battery – a typical car battery

(b) Secondary cells The important difference between a primary cell and a secondary cell is that secondary cells can be safely recharged. The commonest example of a secondary cell is the standard car battery or lead/acid battery.

(i) Lead–acid batteries These batteries consist of cells made up from a series of positive and negative lead electrodes and an electrolyte of dilute sulphuric acid. The electrodes are called *plates*.

'Formed plates' are created during manufacture by repeated charging and discharging of the cells. The negative plate becomes coated with 'spongy lead' and the positive plate with lead peroxide.

'Pasted plates' are manufactured by pressing in sulphuric acid and red lead. A combination of formed and pasted plates is used in large capacity cells. When discharged, the material of both plates combines with the acid to form lead sulphate. Recharging the battery reverses the process. A simple diagram of the lead–acid battery is shown in Figure 16.2.

When current is drawn from the cell the active chemicals on the positive plate expand and the plate tends to distort, especially under heavy loads. Some measure of protection is afforded by having each positive plate adjacent to two negative plates as shown in Figure 16.2.

Care of lead–acid batteries

Since the battery contains sulphuric acid, the case should be handled with care; sulphuric acid fumes are noxious and should not be breathed in. Charging should therefore be carried out in a well-ventilated area, and the environment chosen such that any acid splashes do not cause any damage to the surroundings.

As the battery delivers a current, the plates become coated with lead sulphate, which weakens the acid. A measure of this weakening can be obtained by checking the specific gravity (SG) of the electrolyte – in this case, the sulphuric acid – which can be done by using a hydrometer as shown in Figure 16.3.

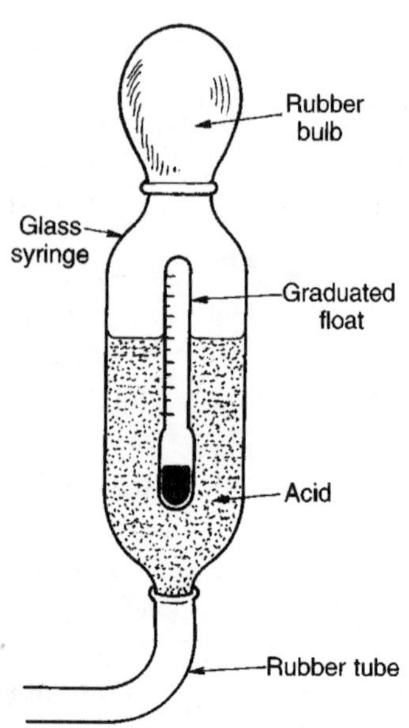

Figure 16.3 A hydrometer

The hydrometer is a graduated tube, weighted at one end. In water, it would just float and measure 1. The more dense the liquid in which the hydrometer floats, the higher it rises, and hence the higher the specific gravity. As the rubber bulb is squeezed with the rubber tube in the liquid, air is forced out. When the pressure is relaxed, liquid is 'sucked' into the glass syringe. The reading on the float can then be taken and some possibilities are shown in Table 16.1.

Table 16.1 SG and charge for a lead–acid battery

SG	*Percentage of charge*
1.28	100
1.25	75
1.22	50
1.19	25
1.16	Fully discharged

Plate colour

In a healthy cell, the positive plate is brown and the negative plate is slate-grey. When discharged, both plates look black.

General maintenance procedures

- When preparing an electrolyte, always add acid to the water, not water to the acid.
- Ensure that any maintenance is carried out in a well-ventilated area.
- Ensure that any naked flames are not used near the cells.
- Where cells remain unused for any length of time the electrolyte should be left in and a periodic charge given to keep the cell healthy until needed again.
- A lead–acid battery should never be left in an uncharged state as a layer of whitish lead sulphate will form on the plates, which will increase the internal resistance and reduce the capacity of the cell. This process is known as sulphation of the plates.
- Terminals should be coated with petroleum jelly in order to prevent corrosion.

(ii) Nickel–alkaline cells There are two types, nickel–iron ('NIFE' cell) and nickel–cadmium. Since 'NIFE' cells are no longer in common use there is no point in considering them here. The nickel–cadmium types (in one form or another) are those currently available in High Street stores under various brand names as rechargeable cells. The anode is composed of nickel hydroxide and the cathode from cadmium (with traces of iron). The electrolyte is potassium hydroxide and the nominal voltage is 1.3 V. These batteries require no maintenance at all other than the appropriate recharging. Table 16.2 lists the advantages and disadvantages of lead–acid and alkaline batteries.

Table 16.2 Lead–acid and alkaline batteries compared

Cell	*Advantages*	*Disadvantages*
Lead–acid	Inexpensive	Fragile
	Working life of between 4 and 6 years	Temperature affects efficiency
	High discharge voltage	Self-discharges when not in use
	Materials used in manufacture are easy to obtain	Requires regular maintenance
Alkaline	Very robust	Very expensive
	Retains its charge when not in use	
	Needs little or no maintenance	

The three effects of an electric current

1 Heating effect

The movements of the electrons within the conductor, which is the flow of electric current, causes an increase in the temperature of the conductor. The amount of heat generated by this current flow depends upon the type and dimensions of the conductor and the quantity of current flowing. By changing the variables, a conductor may be operated hot and used as the heating element of a fire, or be operated cool and used as an electrical conductor.

The heating effect of an electric current is also the principle upon which a fuse gives protection to a circuit. The fuse element is made of a metal with a low melting point and forms a part of the electric circuit. If an excessive current flows, the fuse element melts, breaking the circuit.

2 Magnetic effect

Whenever a current flows in a conductor a magnetic field is set up around the conductor in circular paths. The magnetic field's strength increases with an increase in current and collapses if the current is switched off.

A current-carrying conductor wound into a solenoid produces a magnetic field very much like that of a permanent magnet of similar dimensions. A metal core placed inside the solenoid produces an electromagnet when current is flowing; when the current is interrupted, the magnetic field collapses. The magnetic effect of an electric current is the principle upon which electric bells, relays, moving coil meters, motors and generators work. See also Chapter 4.

3 Chemical effect

When an electric current flows through a conducting liquid, the liquid is separated into its chemical parts. The conductors which make contact with the liquid are called the anode and cathode. The liquid itself is called the electrolyte and the process is called electrolysis.

The electrolytic effect used for the deposition of one metal on top of another is an example of the chemical effect of a current. Copper, nickel and chromium plating are all achieved in this way. In electrolytic capacitors, a thin film of aluminium is deposited onto one of the plates.

Electrolysis is the basis of operation of simple cells. A battery is made up of two or more cells as we have seen. Batteries have many useful applications, the most obvious of which is to provide portable electric power.

Questions

1 State:

(a) The type of battery usually used for starting purposes in motor vehicles.
(b) The polariser in an alkaline–manganese cell.
(c) The unloaded emf of an unused Leclanché cell.
(d) The positive active material in a fully charged lead–acid cell.

2 What is:

(a) The anode of a mercury cell made of?
(b) The colour of both plates in a discharged lead–acid cell?
(c) The colour of the positive plate in a healthy lead–acid cell?
(d) The colour of the negative plate in a healthy lead–acid cell?

3 When current is drawn from a typical lead–acid battery, the active chemicals on the positive plate expand and tend to distort it. What design measures are taken in order to afford some protection against this?

4 Draw a diagram of a hydrometer, explaining how it is used and what values you would expect when the liquid is:

(a) water
(b) sulphuric acid
(c) turpentine.

5 Compare the advantages and disadvantages of lead–acid and alkaline secondary cells.

6 List four things that will affect the resistance of a conductor.

7 State the electrical effect responsible for the production of each of the following:

(a) an electric bell
(b) an electric cell
(c) a fuse blowing
(d) an electric motor
(e) the temperature coefficient of resistance
(f) nickel plating
(g) an incandescent electric lamp
(h) an 'AVO' meter
(i) an electric generator.

Multiple choice questions

1 The emf of a cell or battery is

A the same as its potential difference
B the unloaded terminal voltage
C the loaded terminal voltage
D the p.d. across the internal resistance

2 The main difference between primary and secondary cells is

A primary cells are much cheaper
B secondary cells can be recharged
C primary cells can be connected in series
D secondary cells can be wired in parallel

17 The binary system

Digital computers

Modern computers are digital computers. Digital systems operate in a limited number of finite states, usually only two, as is the case with digital computers. Let's consider exactly what is meant by the word 'digital'. Take the example of a clock used to tell the time. In Figure 17.1 each clock shows 5.00, but what will each show next? The digital clock will show 17.01, but the analogue clock, in the meantime, will show all the seconds in between, and we could define hundreds or even millions of different positions in between if we used a microscope to see them. Whether we see them or not, we know that those intermediate stages exist and that the 'minute hand' must pass through them all. The digital display *jumps* directly from zero to one and *there is nothing in between.*

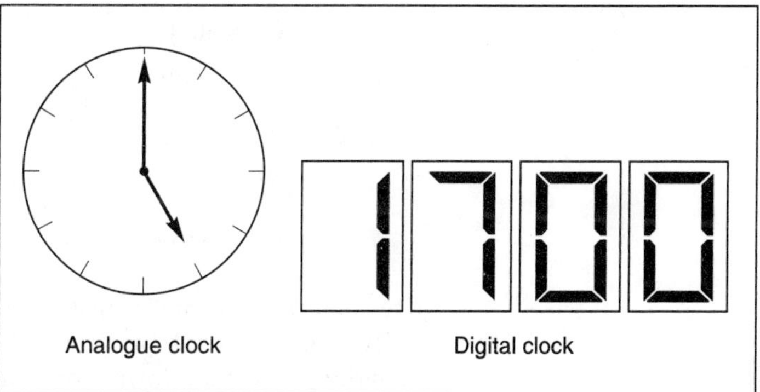

Figure 17.1 Both clocks show 5 o'clock, but what will each show next?

Digital computers operate using tiny transistor switches which are either ON or OFF. Hence, computers are digital devices utilising only two possible states. In digital electronics we refer to these states as logic 1 (ON) and logic 0 (OFF).

A system which has only two states and therefore two 'numbers' (0 and 1) would seem to have a very limited capability, but it will soon be shown that this is not true. Since the digital computer has only two numbers to work with, we have to communicate with the computer only with those numbers. The numbers are simply 1 and 0, and we communicate with the BINARY SYSTEM.

Binary mathematics

Before considering the binary system let's have a quick review of the familiar decimal system. It is called 'decimal' because there are ten digits available, 0 to 9. All other numbers in the system are made up of combinations of these. (Dec means 10 as in decade – 10 years.) A number like 255 really means:

100s	10s	Units
2	5	5

The number 255 is instantly recognisable but it could be worked out as follows:

$$
\begin{aligned}
2 \times 100 &= 200 \\
5 \times 10 &= 50 \\
5 \times 1 &= 5 \\
\end{aligned}
$$

then we add: $\overline{255}$

In 'index' notation (powers of 10) the column headings are:

10^3	10^2	10^1	10^0
(1000s)	(100s)	(10s)	(Units)
4	5	2	7

i.e.

$$
\begin{aligned}
4 \times 10^3 &= 4000 \\
5 \times 10^2 &= 500 \\
2 \times 10^1 &= 20 \\
7 \times 10^0 &= 7 \\
&\overline{4527}
\end{aligned}
$$

(Remember, incidentally, that $x^0 = 1$, whatever the value of x.)

By using decimal numbers, we have illustrated the precise format of writing numbers down, something which we are probably no longer conscious of. As we have seen, however, computers only work with binary numbers – and so must we. In the binary system the column headings are in powers of 2 as follows:

2^7	2^6	2^5	2^4	2^3	2^2	2^1	2^0
(128)	(64)	(32)	(16)	(8)	(4)	(2)	(1)

In all number systems, the column headings are in powers of the number of digits in the system. In decimal there are ten digits (0–9) so the column headings are in powers of 10. In binary, there are two digits (0 and 1) so the column headings are in powers of 2. A list of decimal numbers with their binary equivalents is given in Table 17.1.

Converting from decimal to binary

Say we want to convert the decimal number 14 into its binary equivalent. This can be done by inspection, which, like most things, becomes easier with experience. We need to list the binary 'column headings' first, note

Table 17.1 Decimal numbers 0 to 64 with their binary equivalents. Note how the convention of writing the numbers in groups of four is adhered to. Frequently, the leading zeros are also written down, e.g. 62D would be written as 0011 1110.

Decimal	*Binary*	*Decimal*	*Binary*
0	0	33	10 0001
1	1	34	10 0010
2	10	35	10 0011
3	11	36	10 0100
4	100	37	10 0101
5	101	38	10 0110
6	110	39	10 0111
7	111	40	10 1000
8	1000	41	10 1001
9	1001	42	10 1010
10	1010	43	10 1011
11	1011	44	10 1100
12	1100	45	10 1101
13	1101	46	10 1110
14	1110	47	10 1111
15	1111	48	11 0000
16	1 0000	49	11 0001
17	1 0001	50	11 0010
18	1 0010	51	11 0011
19	1 0011	52	11 0100
20	1 0100	53	11 0101
21	1 0101	54	11 0110
22	1 0110	55	11 0111
23	1 0111	56	11 1000
24	1 1000	57	11 1001
25	1 1001	58	11 1010
26	1 1010	59	11 1011
27	1 1011	60	11 1100
28	1 1100	61	11 1101
29	1 1101	62	11 1110
30	1 1110	63	11 1111
31	1 1111	64	100 0000
32	10 0000		

the largest number which will go into the number to be converted, note the remainder and then repeat the process.

We begin by writing down the column headings:

2^7	2^6	2^5	2^4	2^3	2^2	2^1	2^0
(128)	(64)	(32)	(16)	(8)	(4)	(2)	(1)

16 is obviously too large (again, experience will quickly lead us to write down only the columns we will actually need); so we start with 8 and write a '1' below it. We have therefore represented 8 by putting a '1' in that column, this leaves 6 (14 – 8) which is one 4 and, finally, one 2. Continuing like this it is easy to see that:

14 = (1 × 8), (1 × 4) and (1 × 2)

Hence:	2^3	2^2	2^1	2^0
	1	1	1	0

So, 14 (decimal) is equal to 1110 (binary).

There are several conventions, but we shall use simply the letter D to indicate a decimal number, and the letter B to indicate a binary, i.e. 14D = 1110B.

Now let's convert 63D into binary:

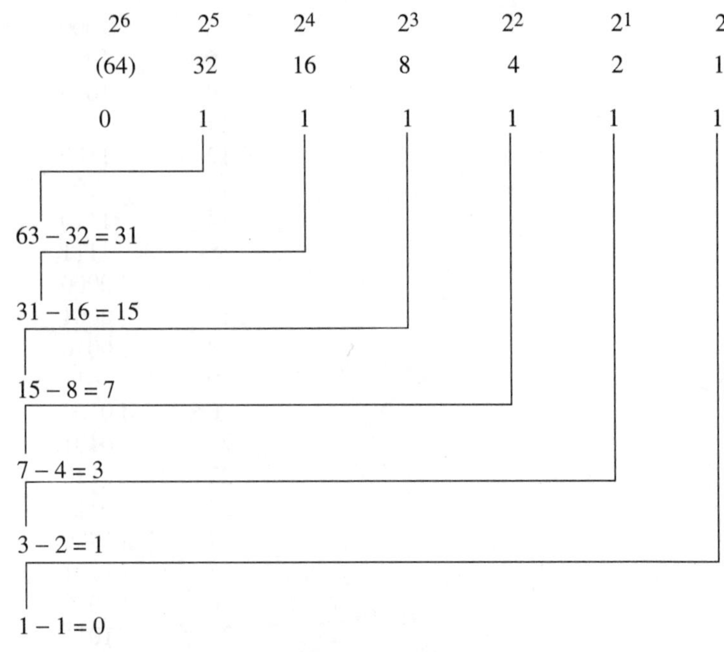

i.e. **63D = 11 1111B**

The answers to two other problems are given below; try these for yourself until you find the correct solution:

92D = 101 1100   245D = 1111 0101

There is another (probably better) method of making these conversions. This is known as successive division by 2. As an example, let's use this method to convert 57D to binary. To do this by successive division by 2, divide 57 by 2, obtaining 28 and a remainder of 1. Note these two numbers as shown and then continue dividing by 2:

57/2 = 28	1	remainder values
28/2 = 14	0	
14/2 = 7	0	
7/2 = 3	1	
3/2 = 1	1	
1/2 = 0	1	

When the divisions have been completed, write down the remainder values in reverse order.

Therefore: 57D = 11 1001B. This is because the remainder values have been taken in reverse order and written down to form the binary equivalent.

Note the next example:

$$92/2 = 46 \quad 0$$
$$46/2 = 23 \quad 0$$
$$23/2 = 11 \quad 1$$
$$11/2 = 5 \quad 1$$
$$5/2 = 2 \quad 1$$
$$2/2 = 1 \quad 0$$
$$1/2 = 0 \quad 1$$

Therefore: 92D = 101 1100B

The answers to two other problems are given below; try these for yourself until you obtain the correct solution:

179D = 1011 0011B 133D = 1000 0101

Converting from binary to decimal

The easiest way of doing this is to write down the column headings as powers of 2 and/or the numbers they represent. Then write down the binary number below and add up the columns. The following examples should make this clear.

Convert 11001B to decimal:

2^4	2^3	2^2	2^1	2^0
(16)	(8)	(4)	(2)	(1)
1	1	0	0	1

$= 16 + 8 + 1 = 25$
i.e. 11001B = 25D

Convert 11010B to decimal:

16	8	4	2	1
1	1	0	1	0

$= 16 + 8 + 2 = 626$
i.e. 11010B = 26D

Convert 1110 1011B to decimal:

128	64	32	16	8	4	2	1
1	1	1	0	1	0	1	1

$= 128 + 64 + 32 + 8 + 2 + 1 = 235$
i.e. 1110 1011B = 235D

Simple binary arithmetic

Adding and subtracting binary numbers is similar to decimal arithmetic. If anything it is rather easier; all you need to know is the following:

$0 + 1 = 1$; $1 + 1 = 10$ and $1 + 1 + 1 = 11$

$1 - 0 = 1$; $10 - 1 = 1$ and $11 - 1 = 10$

The following examples should make the process clear:

```
1   101 +      2   1101 +      3    1011 +
    110            1111            11011
    ----           -----           ------
    1011           11100           100110
```

When subtracting, if you need to take 1 from 0, 'borrow' a '10' and note that 10 – 1 = 1. For example, subtract 1101 from 10011:

```
10011 –
 1101
 ----
 0110
```

1 – 1 = 0, and 1 – 0 = 1.
0 – 1 can't be done, so we need to borrow a 10. But there isn't one, so we have to progress along the line until we find one.

The leftmost digit therefore disappears, making the next 0 into 10. This in turn is reduced to 1 (since 10 – 1 = 1) when 1 is borrowed from it. The difference then becomes:

```
0  1  10  1  1
   1   1  0  1
--------------
   0   1  1  0
```

There is also something to be said for converting the binary numbers into decimal, doing the arithmetic and then converting back again.

These numbers are fairly simple, but difficult to recognise and handle; for example, what is 1110 1011B in decimal? And what about a really large number like 64 424? This turns out to be 1111 1011 1010 1000 in binary! Of course, we now know how the conversions are made and we also know what a boring, time-consuming and tedious process it is. Furthermore, we have also learned that this is the only language that a microprocessor understands and we're stuck with it. Imagine having to program a computer using only binary codes. In fact, this is how it used to be done and apart from the disadvantages just mentioned, the problem of avoiding errors was enormous. Clearly a better system of representing binary numbers is required. Such a system exists in the HEXADECIMAL CODE but detailed knowledge of it is not required at this level.

Binary coded decimal (BCD) numbers are of great use and significance in digital electronics and these are discussed in some detail in Chapter 9.

Questions

1 Convert the following binary numbers into decimal:

(a) 11 (b) 101 (c) 1100 (d) 1111
(e) 1100 0011 (f) 0110 0111 (g) 1110 1111

2 Convert the following decimal numbers into binary:

(a) 27 (b) 45 (c) 16 (d) 83
(e) 245 (f) 77 (g) 392 (h) 415
(i) 42 (j) 249 (k) 511 (l) 1000

3 Add the following binary numbers together:

(a) 101 + 110
(b) 1101 + 1111
(c) 1011 + 11011
(d) 111 + 10011
(e) 1001 + 1101
(f) 11011 + 11011
(g) 11110 + 101101
(h) 110111 + 101
(i) 1011 + 1011100
(j) 100011 + 1010101

4 Subtract the following binary numbers from each other:

(a) 111 – 101
(b) 1111 – 1001
(c) 10011 – 1101
(d) 11001 – 1111
(e) 101010 – 10101
(f) 10011 – 1001
(g) 10110 – 1111
(h) 101011 – 111
(i) 100001 – 11101
(j) 1001101 – 11110

18 Health and safety

Only limited knowledge of the Health and Safety at Work etc. Act 1974 is required; under this Act an employee at work must 'take reasonable care for the health and safety of himself and others'. These notes will help you to recognise and indicate potential health and safety hazards in the handling and use of materials and equipment in work situations.

Some important points include the correct wiring of a 13 A plug top, use of the correct size fuse, what to do if you discover a person suffering from an electric shock, how to treat burns, and what kind of extinguisher to use on electrical fires.

> Under the Health and Safety at Work etc. Act 1974, an employee at work must:
>
> 'take reasonable care for the health and safety of himself and others'.

Figure 18.1 shows a TV engineer repairing an AC/DC television (AC/DC means that the equipment will work on 240 V DC as well as 240 V AC – so the equipment contains a rectifying system but no mains transformer). One side of the mains (which should be the neutral) is connected directly to the metal chassis of the TV. In this case, the TV has been incorrectly wired so that the bare chassis is connected to the live side of the mains.

While touching the chassis with one hand, the engineer leans against and touches an earthed water pipe with the other hand. The mains current travels through the man's body and he is unable to let go. This could easily result in death. Impossible? There is a story of a technician in a school who was repairing such an AC/DC TV. Because TV aerial outlets are required throughout the school, a distribution amplifier was used and it was earthed. This meant that the coaxial plug on every TV fly lead was also earthed. The technician had his left hand on the TV chassis which had inadvertently been connected to the live side of the mains. As he plugged the aerial lead into the socket on the wall, his right hand became earthed via the distribution system and he received a potentially lethal shock from one hand to the other, across his chest – one of the worst types of electric shock – and he could not let go.

There was no one else in the building at the time and the technician could easily have died. He was saved because he was working on a stool; the shock caused him to fall backwards and, in so doing, the aerial fly lead was pulled out of the wall socket, disconnecting the earth connection. I know this is true; I was that technician!

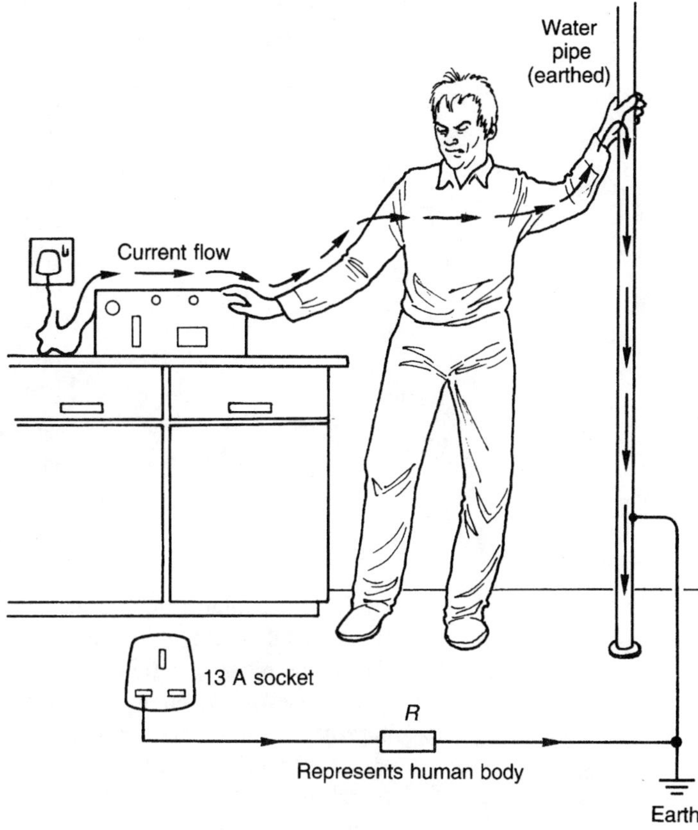

Figure 18.1 A TV engineer receiving an electric shock

No one should work alone in a workshop or laboratory. In case of electric shock another person being around to switch off the current or render assistance could be a life saver.

Dangers of electricity

Human life depends upon the correct operation of the heart. This is a muscle which acts as a pump. Associated with its operation, as with any other muscle, is a very small electric signal. Because the body contains fluids it will conduct electricity. These applied currents (it is the current not the voltage which matters) interfere with the normal operation of the muscles and start to control their operation. The effects of electric shock are summarised in Table 18.1.

Remember that 500 mA is only 0.5 amp. Even this relatively small current can lead to unconsciousness or even death. The need to be extra careful with electricity is therefore obvious. For this reason, isolating transformers which limit the current were always fitted in laboratories and workshops. Nowadays, earth leakage circuit breakers (ELCBs) and residual current devices (RCDs) are more frequently employed and much

Table 18.1 The physiological effect of current through the body

Current (mA) *at* 50 Hz	*Physiological effect*
0–1	Range up to threshold of perception.
1–15	Range up to cramp. At upper limit unable to release grip.
15–30	Cramp-like contractions of the arms. Limit of tolerance. Rise in blood pressure; difficulty in breathing.
30–50	Powerful cramp effect. Rise in blood pressure. Unconsciousness. Heart irregularities.
50–500	Unconsciousness; current marks.
Over 500	Unconsciousness; current marks; burns. Heart irregularities; could lead to cardiac arrest; death.

equipment contains double insulation. However, whatever safety precautions are taken, it is ultimately the good practice and care of the individual that is the most important.

Because of the obvious importance of the need for being particularly careful with electrical equipment, a written source of good practice in electrical work is available. This is called the *IEEE Regulations for Electrical Installations.* Of particular relevance to electronics engineers is the Health and Safety at Work etc. Act 1974.

Protection is devised so that, under normal working conditions, it is not possible to touch a naked live wire. Where such wires may come into contact with metal surfaces which may be exposed, such metal surfaces should be earthed.

Electrical equipment becomes more dangerous when wet, so that working in wet or damp conditions is extremely hazardous. If our bodies are earthed, an electric shock is more likely to be lethal, so that working outside is potentially more dangerous. Every year, people die needlessly in their gardens, mowing the lawn or clipping the hedge. Good practice and the use of an ELCB or RCD could save many lives.

A casualty in contact with live equipment should not be touched. The first thing to do is to turn the mains off as soon as possible. If this cannot be done immediately, the victim should be dragged away using a scarf or similar item of clothing or pushed away with something like a broom handle.

If the patient is unconscious, check whether or not he or she is still breathing. If breathing has stopped or if the face is blue and there is breathing difficulty, lift the chin and tilt the head back. If breathing does not start or improve, remove any obstruction from the mouth or throat (e.g. dentures) and commence mouth–to–mouth resuscitation or other artificial respiration.

Burns

Burns can be caused through the heating effect of electricity when a high current is available. High voltage sparks also cause burns, even when the current is very low. Burns caused by fire or electricity should be cooled using clean, running water, unless the damage is too severe and a clean or sterile, dry dressing applied. Creams and lotions should *not* be used. If anything is sticking to the burnt area it should not be removed except by qualified medics.

Electrical safety – a summary

Safety area	*Important points*
Means of isolating supplies and testing that circuits are dead	Knock off main isolator; remove appliance plug; check with a neon or AVO.
Dangers associated with work on live systems	Live/earth through the body can kill, unable to let go.
Safe use of manual and powered tools	Keep mains lead away from cutting edges; earth; RCD.
Cables and leads, plugs and socket outlets	White cables best; check no frayed edges; 13 A plugs wired correctly.
Importance of competent workmanship and correct connections	Danger of earth wires coming adrift.
Need for regular inspection of tools and equipment	Make regular visual checks; use an electronic tester.
EMERGENCIES	Know where the nearest mains isolator is; know where the nearest telephone is.

General safety

Safety area	*Important points*
Accident prevention; dangers of careless or untidy working	'Good housekeeping'.
Use of safety and protective clothing; protection of face, eyes and hands	Gloves, goggles, shields.
Soldering and desoldering	Correct use of equipment.
Lifting and handling	Correct posture when lifting.

Fire safety

Safety area	*Important points*
Fire prevention; conditions for combustion	Keep fire doors closed; Heat, oxygen, fuel.
Methods of dealing with different types of fire	Water on *all* fires *except* electrical fires and flammable liquids; use foam or CO_2.

contd.

Safety area	*Important points*
Types of fire extinguisher and their appropriate uses	Water, foam, dry powder CO_2, blankets, dry sand.
Dangers from toxic fumes and smoke	Beware of certain plastics and use fire doors.

First aid

Safety area	*Important points*
Extent of first aid to be offered in the event of burns	Douse in cold water; take no other action. There is no such thing as 'burn ointment'.
Dealing with cuts, contact with irritant material or subjection to toxic fumes and smoke	Remove from danger area; apply pressure to cuts unless glass present.
Importance of seeking qualified assistance	Do not give any stimulants.
ELECTRIC SHOCK	
Symptoms and effects of electric shock	See earlier notes.
Need for immediate action to remove victim from contact and dangers involved	*Switch off* at mains; *do not* touch the victim.
Methods of resuscitation; checking for other injuries	'Kiss of life', etc. Do not move victim.
Immediate treatment of burns	Apply cold water; do not cover; seek medical help.

Some modern consumer units have an automatic main switch which will switch off all the circuits in the house in a very short time (typically 25 ms) if an earth leakage current of more than about 25 mA occurs. Any current flowing into the earth means that a fault has occurred. In the absence of a fault, the current flowing in live and neutral wires will be the same. An earth leakage circuit breaker (ELCB) will detect any imbalance between the currents in the two wires and switch off the current. A residual current device (RCD) is simply another type of ELCB.

Electrical safety in industry

Many companies and organisations have specifically authorised personnel to replace fuses, plugs and wires, etc., and to carry out general electrical maintenance work. THIS IS VERY IMPORTANT:

1 *Competence*: in a large organisation, an individual's ability may not be recognised by the management. Specific competences, generally certificated, are required.

2 *Legal requirement*: in case of injury (or death) to an employee or visitor, the company may be sued for negligence, etc.
3 *Fire hazards*: only properly qualified personnel will be able to spot potential fire hazards.
4 *Union demands*: trade unions may take the view that if personnel employed and paid to do a specific job find someone else doing it for them, their jobs may be threatened. Therefore only electrical maintenance personnel or electricians should do that job.
5 *Insurance*: in the event of an accident, injury or fire, etc., insurance companies may not pay up if repairs, maintenance and servicing are not carried out by properly qualified personnel.
6 *Guarantees on equipment and installations*: these may be invalidated if equipment and installations are not properly maintained or if they have been tampered with.
7 *Preventative maintenance*: qualified personnel will be able (and be required) to spot other defects in a system, equipment or installation and correct any defects before they become a hazard.

ALWAYS ALLOW ELECTRICAL MAINTENANCE TO BE CARRIED OUT BY QUALIFIED PERSONNEL

The notes in this section are intended as a guide only. It is essential that workers should have a good knowledge of first aid or know the name and location of someone who is suitably qualified. Similarly, the location of first aid boxes should be common knowledge and someone appointed to ensure that they are always adequately stocked. Modern first aid boxes carry the BS 5378 first aid sign which consists of a white cross on a green background.

13 A plug wiring

By law, all new electrical appliances are now fitted with wires coloured in the European code: brown for live, blue for neutral and yellow/green stripe for the earth, terminated with an approved 13 A plug. The old British Code (red – live, black – neutral, green – earth) is no longer used and it is now illegal to offer for sale any electrical equipment with the old colour-coded cable attached to it.

The live and neutral wires carry the current, the earth is added for safety. Some appliances only have two wires (live and neutral) and this is usually because the equipment is double insulated (e.g. hair driers) or it is AC/DC equipment which does not contain a transformer (rare nowadays).

Connecting a cable to a 3-pin plug

1 Figure 18.2 shows one type of 13 A plug, there are many approved variations so you will have to adapt these notes according to the plug

being used. The first thing to do is to remove the back of the plug and estimate the length of wire required to reach the large earth pin. The outer sheath of the cable should be firmly held by the cable grip such that 0.5 cm (about $\frac{1}{4}''$) of the sheath is visible inside the plug.

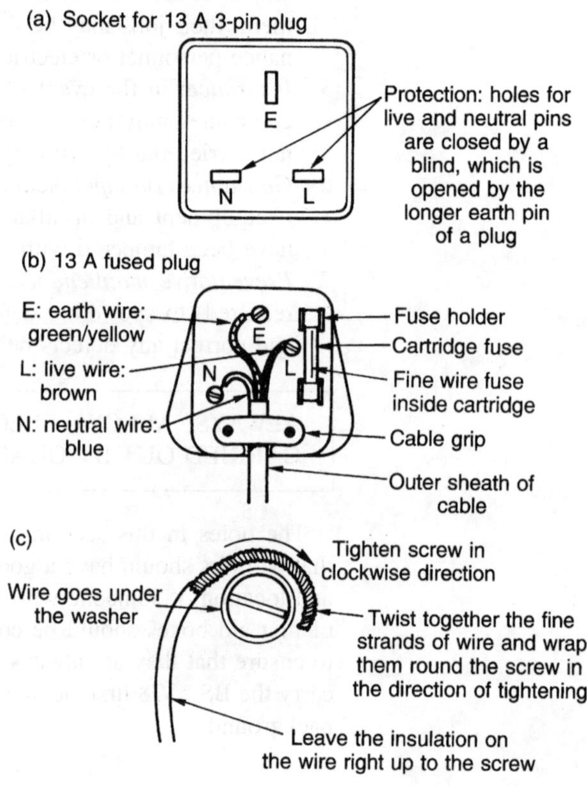

Figure 18.2 13 A plug wiring details

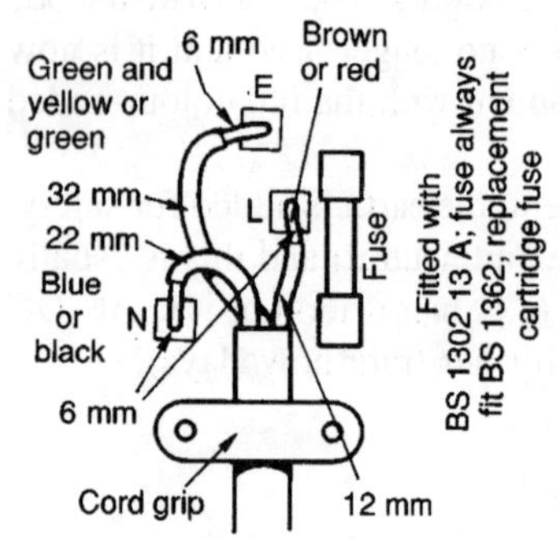

Figure 18.3 13 A plug detail

2 The outer sheath should be cut away carefully to avoid damaging the coloured insulation of the three wires inside.
3 The live and neutral wires should now be cut to the appropriate lengths and about 1 cm of insulation stripped from the ends of all three wires. Particular care should be taken not to cut off any of the fine strands of wire and to ensure that there is sufficient insulation remaining to protect the wire right up to the pins of the plug.
4 The fine strands of wire should be twisted together so that no stray strands are left loose inside the plug.
5 Figure 18.3 shows the precise measurements recommended for a particular plug. This may be used as a guide although it is not necessary to measure the wires.
6 If the plug has wrap-around screw terminals, the ends of each wire should be bent round in the direction of tightening of the screw.

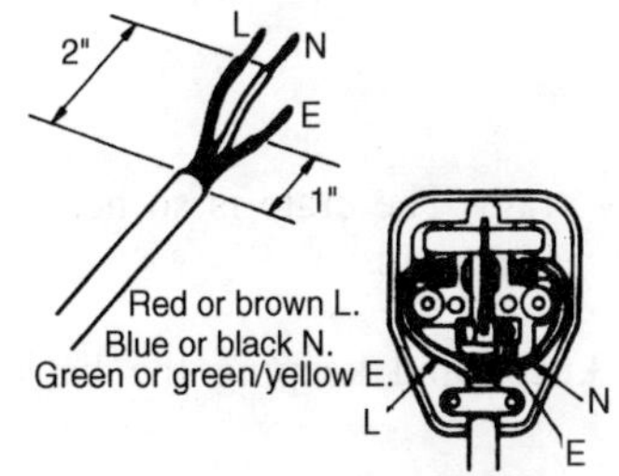

Figure 18.4 13 A trailing socket

7 Ensure that the correct colour code has been used, the cable grip is so tight that the cable cannot be pulled out and the correct value fuse is fitted. If all is well, the cover of the plug may be replaced.

13 A Trailing sockets

The procedure for fitting a 13 A trailing socket is similar to that described for the 13 A plug – but remember to put the cable through the outer casing first!

Electronic projects

When constructing electronic projects which have to be connected to the mains supply, the following guidelines should be observed:

1 The whole unit should be built into an earthed metal box.
2 As well as bolting on an earth connection, the mains earth wire should be directly soldered to the metal box.
3 Double wound transformers should be used.
4 A physical metal barrier should be placed between mains and low voltage sections.
5 The mains lead should be secured by an approved cable cleat.
6 A neon lamp should be connected to the mains input, such that it glows as soon as the mains is connected, that is to say, before any switch the unit may contain.
7 'Star' washers should be used with any fixing screws, especially for any lid or cover which provides access to the circuitry inside.

You may also wish to include double-pole switches, mains and low tension fuses and LED indicators, etc., in addition.

Multiple choice questions

1 The colour used to represent earth on a (British) mains cable is

A black
B green
C yellow
D green/yellow

2 The BS 5378 first aid sign is

A a white cross on a green background
B a green cross on a white background
C a white cross on a red background
D a red cross on a white background

3 When desoldering, it is essential to use

A a soldering iron rated at 100 W or more
B a special desoldering tool

C appropriate eye protection
D appropriate hand protection

4 A technician is asked to move a heavy crate. The crate is so heavy that the technician cannot even slide it. He should

A attempt to lift it by bending the knees and keeping the back straight
B ask a workmate to assist
C ask two workmates to assist
D request mechanical means to move the crate

5 The maximum voltage likely to be encountered in a domestic TV set is

A 250 V
B 2500 V
C 25 kV
D 250 kV

6 Care should be exercised when handling live power connectors in order to avoid

A a lethal electric shock
B an unpleasant shock through the insulation
C the connector breaking
D incorrect insertion

7 An electric heater fails to work in the workshop. The most likely person to be authorised to investigate the fault is

A a test engineer
B a storeman
C an electrician
D the personnel officer

8 A safety sign which is triangular with a yellow background surrounded by a black band is a

A prohibition sign
B warning sign
C mandatory sign
D safe condition sign

9 A safety sign with a blue background and white symbol is a

A prohibition sign
B warning sign
C mandatory sign
D safe condition sign

10 Mains soldering irons should be

A earthed
B fitted with very long leads
C fitted with very short leads
D fitted with iron bits

11 The greatest danger of testing equipment on load is

A very high currents will be flowing in the equipment
B inadequate test results will be obtained
C the internal fuses are most likely to blow
D earth connections need to be removed

12 A technician accidentally drops his soldering iron. He should

A try to catch it
B let if fall and pick it up again carefully as soon as possible
C let it fall and leave it where it lands until it has cooled down
D let it fall and report the matter to the supervisor

13 Multicore tin/lead solder used in electronics melts at around

A 120°C
B 230°C
C 500°C
D 1000°C

14 An RCD is a

A resistance/current device
B restricting current device
C residual current device
D recharging device

15 An ELCB is an

A earth leakage circuit breaker
B earth line connection bond
C electricity local circuit board
D electronic low current breaker

Part Four
Practical Work

19 Introductory practical work

Breadboarding

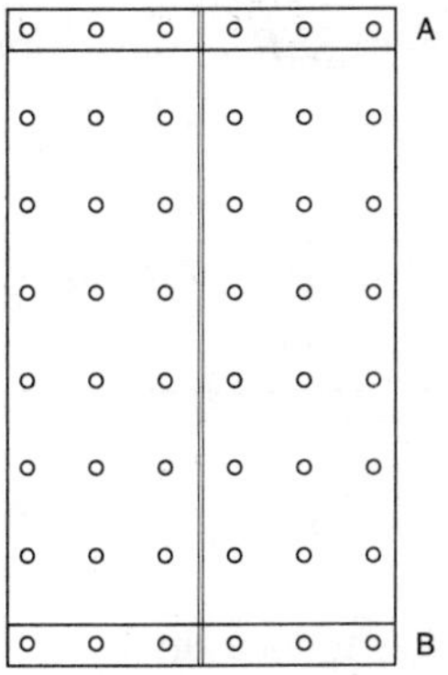

It is said that the term 'breadboarding' originated at a time when hobbyists and other technically interested people were first beginning to assemble radios and other electronic devices at home. In those early days (the 1920s and 1930s perhaps) electronic hardware and components were not as freely available as they are now. Accordingly, a few inventive home constructors acquired a breadboard, hammered some nails or screws into it and wired the electronic components in between.

The story may be apocryphal, but the term 'breadboarding' is now widely used to describe a method of putting together a prototype electronic circuit. Although strictly speaking it is not a breadboarding system (the components are all mounted on 'carriers'), the one which is now describe is called 'Locktronics'.

This system consists of a panel on which are mounted several metal posts, arranged symmetrically and between which electronic components mounted on 'carriers' may be arranged. It is thus an easy matter to build up many basic electronic circuits quite simply and without the need for soldering. It also means that the circuits can be dismantled easily afterwards and all the equipment used again and again.

The layout of a Locktronics panel is shown below in Figure 19.1.

A

B

Figure 19.1 A Locktronics panel. All the pins at A are connected together, and all the pins at B are connected together; although A and B are separate

The use of meters

This introductory section will give you some practice in using electronic meters to measure voltage and current. It also shows how meters can distort the voltages they are trying to measure; this is because of their INTERNAL RESISTANCE.

Many discrete (separate) meters are available for measuring voltage and current. There are those used in laboratories for making measurements in various experiments and there are those which can be mounted permanently in electronic equipment, notably in power supplies. Most people will be familiar with the ammeter used in a battery charger, for example.

For measurements in electronics servicing and repair, however, it is usual to use a MULTIMETER – an instrument with switchable ranges which can measure voltage and current, on either AC or DC, plus various resistance measurements. Multimeters fall into two distinct categories: analogue and digital. Of the analogue type, the most famous is the AVO meter (short for amps, volts and ohms).

There are also many electronic meters usually called either digital voltmeters (DVMs) or digital multimeters (DMMs). The internal resistance of AVO type analogue meters and DVMs varies considerably and this assignment will reveal some of the implications when making certain measurements.

Notes

1 *Always* switch off meters when not in use:

(i) DVM or DMM (digital) – saves the battery
(ii) AVO (analogue) – protects the meter movement.

2 *Always* check that the power supply is set to *ZERO* volts before turning it on:

(i) it protects the ON/OFF switch
(ii) it prevents an unexpectedly high voltage damaging the equipment it is connected to.

Assignment 1 – Voltage and current measurements

Connect up the circuit of Figure 19.2 using Locktronics or a similar method.

When your circuit is complete, make the following measurements using (i) a DVM and (ii) an AVO meter. Record your results in a table similar to that given in Table 19.1.

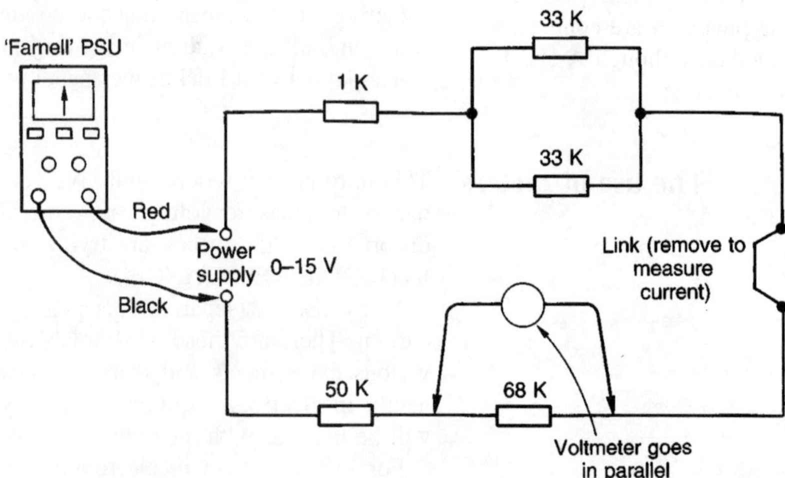

Figure 19.2 Circuit diagram for Assignment 1 – use of meters to measure current, voltage and resistance. Note the use of red and black leads for the power

Note After obtaining the DVM readings, *leave the DVM connected* when the AVO is connected. *Note what happens to the reading on the DVM*. Since both meters are connected to the same point, they *must* both

Table 19.1 Simple voltage measurements using a 15 V power supply

Across resistor	*(a) DVM alone reads*	*(b) AVO reads on* 30 V *range*	*(c) AVO reads on* 100 V *range*
50K			
68K			
33Ks			
1K			
TOTAL (added up)			
Total measured			

read the same value. This value, *taken from the DVM* should be entered in the second column. In the third column, enter the reading obtained when the AVO is switched to the 100 V range.

Think carefully at this stage about what is happening. In column (a) the DVM gives you a certain reading; when the AVO is attached, does the DVM read the same? Is it any different when the AVO is then switched to the 100 V range? Think about it!

Questions

1 Most of the readings taken with a DVM and then an AVO are different. Why should this be so? Why is there a difference between the AVO 30 V range readings and the 100 V range readings?

2 Are there any readings which are very similar? Can you explain this?

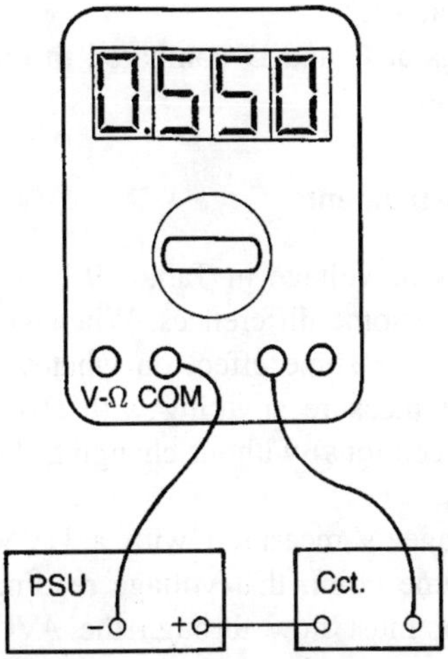

Measuring current

Note that when measuring current (electron flow), the meters are always placed in SERIES with the circuit. This means that the current must go *through* the meter – as shown in Figure 19.3.

When measuring current with a DVM type meter obtain an extra red lead and connect it from the positive of the power supply to the A socket of the meter. Now plug the red lead which previously went to the circuit (from the positive of the power supply) to the socket marked COM on the meter. Set the meter to the highest range in the (grey) 'DCA' (DC amps) section. On many meters, this is the 2 A range.

Turn on the power supply and then move down the ranges until a reading is obtained. This is called 'downranging' and it helps to protect the meter from accidental damage. When using Locktronics you can often measure current by simply removing a link and replacing it with a meter on the amps range.

Using Figure 19.4 as a guide, measure the current in the circuit:

Figure 19.3 Measuring current – the meter leads are in SERIES with the circuit elements

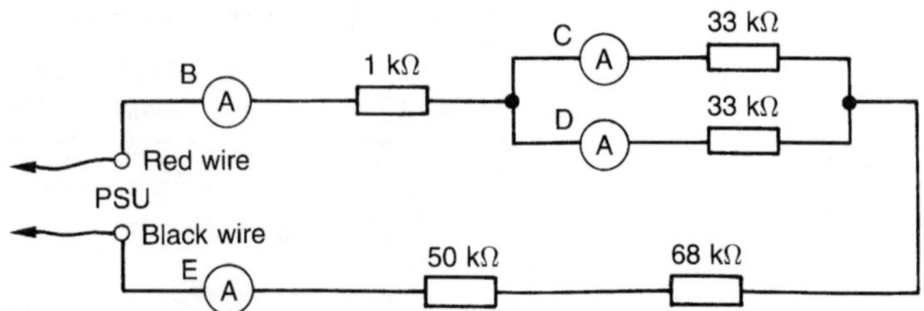

Figure 19.4 Measuring current in the simple circuit

1 where it goes into the circuit at B
2 where it splits at C and at D
3 where it leaves the circuit at E.

Do this when the supply voltage is 15 V and again when it is 30 V. Remember to use the correct POLARITY especially when using analogue meters, and quote the correct units. Record your results as shown in Table 19.2

Table 19.2 Recording current measurements

Supply voltage: 15 V	*Current at*			
	B	C	D	E
Using a DVM Using an AVO				

Questions

1 What do you notice about readings B and E?
2 What do you notice about the readings at C and D compared to the total current?

Internal resistance (or impedance) of instruments

Study the DVM and AVO meter readings of voltage in Table 19.1 you have recently completed. You should notice some differences. Whenever an instrument measures anything, it will have some effect on whatever is being measured. In theory, you cannot measure anything with absolute accuracy (although there are some exceptions) without changing the quantity you are measuring.

You should observe that when a voltage is measured with a DVM and an AVO is then placed across the same point, that voltage reading is *changed.* This is because some current must flow through the AVO (analogue) meter in order for the needle to move and register the voltage it is reading. Analogue meters tend to have a much lower internal resistance than digital meters so the change in voltage is greater.

The AVO meter has an internal resistance of 20 kΩ per volt (20 kΩ/V) so on the 30 V range, the internal resistance of the instrument is $30 \times 20\,k\Omega$

which is 600 kΩ. On a similar range, a DVM might have an internal resistance of 6 MΩ and probably much more. It therefore has a substantially lower effect on the voltage it is reading and (to all intents and purposes) no measurable effect at all.

Figure 19.5 illustrates the effect of the internal resistance of a meter. A potential divider made up of equal resistances (1 kΩ here) produces half the total voltage at its junction (3 V in this case). If a meter with internal resistance of 1 kΩ is connected across the lower resistor, the measured voltage is only 2 V. A meter with a very large internal resistance (10 MΩ in this example) makes hardly any difference at all.

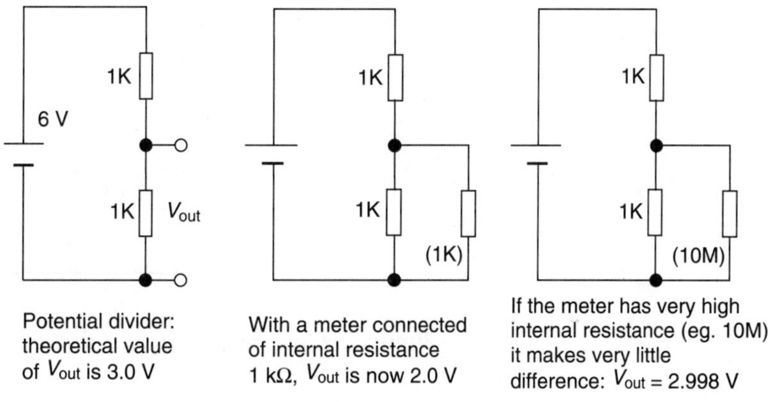

1K
6 V
1K
V_{out}
Potential divider: theoretical value of V_{out} is 3.0 V
1K
1K
(1K)
With a meter connected of internal resistance 1 kΩ, V_{out} is now 2.0 V
1K
1K
(10M)
If the meter has very high internal resistance (eg. 10M) it makes very little difference: V_{out} = 2.998 V

Figure 19.5 Illustrating the effect of meter internal resistance

A typical procedure of the sort being used for NVQ style assessment is given below. This indicates the kind of criteria involved and the particular range of abilities a student would be expected to attain by successfully completing an assignment such as this.

ASSESSMENT

1	Correct use of red and black leads and power supply correctly connected:	☐	☐
2	Circuit built up to specification with adequate layout:	☐	☐
3	Voltage measurements taken correctly and accurately; correct units used consistently:	☐	☐
4	Meter connected correctly for taking current measurements and correct values obtained:	☐	☐
5	Questions answered correctly:	☐	☐

Introducing the transistor

The transistor is a semiconducting device fabricated from silicon. It has three electrodes designated collector, base and emitter. A large current flows from collector to emitter (assuming conventional current flow) but this can be controlled by a much smaller current flowing from base to emitter. See Chapter 9 for further details.

Set up the circuit shown in Figure 19.6 using Locktronics modules.

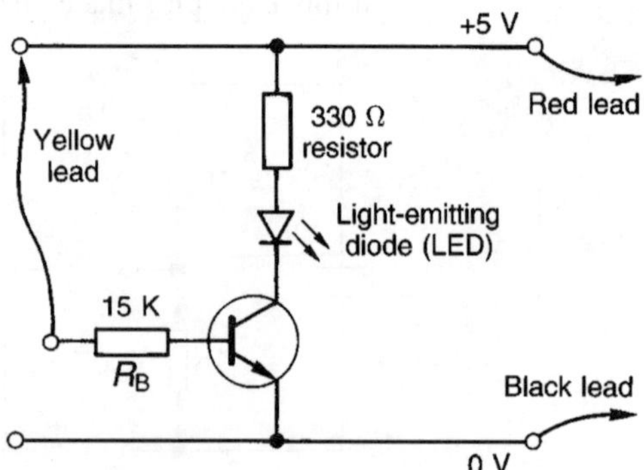

Figure 19.6 The transistor acting as a switch

R_B (about 10 K) is the 'base resistor'. Touch the fly lead on the positive of the supply – the LED should glow indicating that the transistor is turned ON. Connecting the base to 0 V (or simply removing the base connection in this case) turns the transistor switch OFF.

How large a base resistor can you use such that the circuit still works? Further practical work for this section is described in Chapter 9.

Assignment 2 – Construction and testing of an LED flasher

In this assignment we look at what happens if two transistor switches are connected together back to back. The connection is made using electronic components called CAPACITORS; further details are given in Chapter 3.

The circuit shown in Figure 19.7 is correctly called an astable multivibrator. When connected to a power supply (about 5 V is sufficient), the LEDs flash on and off alternately. Connect up the circuit using Locktronics modules or similar. The Locktronics LEDs usually have integral 330R resistors, but do check before using them; use 47 μF cross-coupling capacitors and 10 K resistors. A 5 or 6 V supply is adequate.

Once the circuit is working correctly, make the following changes:

1 Remove C1 and C2 and replace with 0.1 μF capacitors.
2 Remove both LEDs and replace the left hand one with a 1 K resistor.
3 Replace the right hand LED with a small loudspeaker in series with a 1 K resistor.
4 When power is applied, the loudspeaker should emit an audible tone.

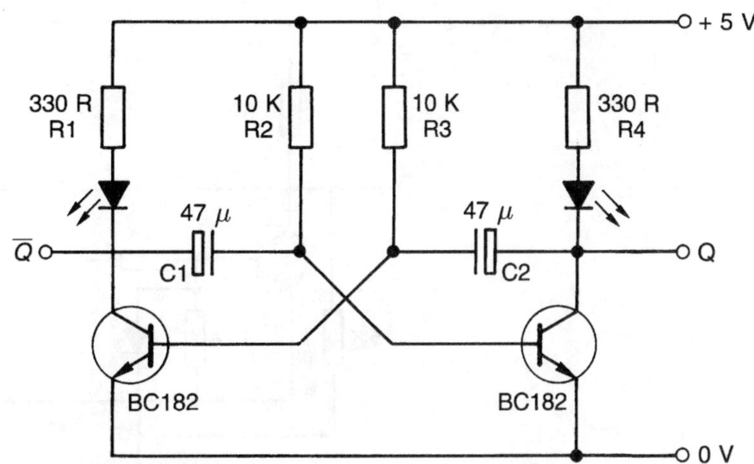

Figure 19.7 An astable multivibrator which flashes LEDs

5 The pitch of the tone can be varied by putting in different values of capacitor, and/or different values of R2 and R3 – try it.
6 The shape of the waveform can be seen by connecting an oscilloscope to the collector of one of the transistors.
7 DO NOT DISCONNECT THIS CIRCUIT AS IT WILL BE REQUIRED TO 'PULSE' THE NEXT CIRCUIT.

ASSESSMENT

1 Correct use of red and black leads and power supply correctly connected:	☐	☐
2 Circuit built up to specification with adequate layout:	☐	☐
3 Circuit functions correctly:	☐	☐
4 Modifications successfully carried out; loudspeaker emits an audible tone:	☐	☐
5 Waveform obtained by correct connection and use of oscilloscope:	☐	☐

Assignment 3 – Construction and testing of a bistable

The astable described in the last assignment is so called because it has no stable state. As long as power is applied, the LEDs continue to flash, or the loudspeaker emits an audible note. The bistable has two stable states and it requires a pulse to 'flip' it from one state to another. For this reason, the circuit is often called a 'flip-flop'. A simple bistable circuit is shown in Figure 19.8.

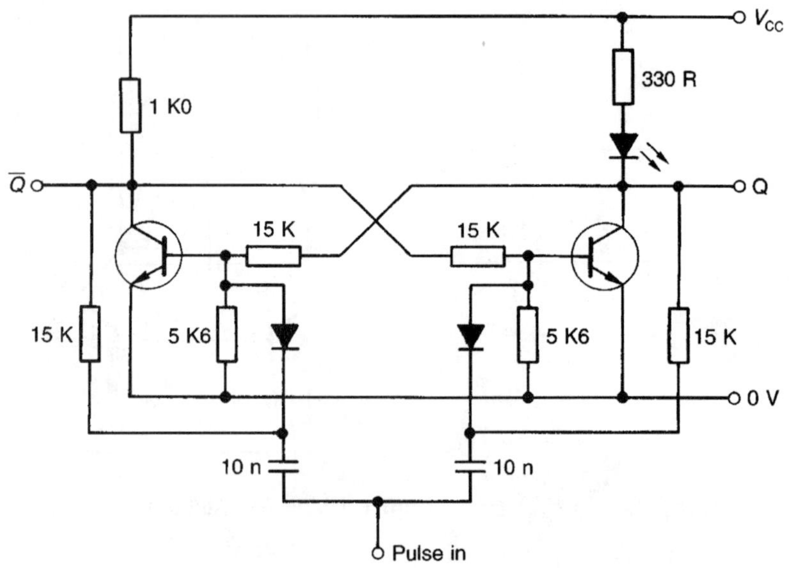

Figure 19.8 A bistable or 'flip-flop' circuit

Connect up the circuit and test it using the astable from the last assignment. Connect the astable, right hand collector output to the 'pulse in' on this circuit. If the LED on the bistable does not flash, try increasing the supply voltage slightly.

The LED should flash at half the rate of the astable.

Assignment 4 – 555 investigations

Obtain the following details about the 555 timer:

1 Maximum and minimum working voltages.
2 Typical current consumption.
3 What is meant by 'sourcing' and 'sinking' and how much current the 555 can source or sink.
4 What the formula is for calculating the operating frequency.
5 The typical component values if the 555 should oscillate at 12.5 kHz.

Make or obtain a 555 timer circuit similar to the one shown in Figure 19.9.

1 Use an oscilloscope and record time-related waveforms at pins 2 and 3. From these waveforms work out the frequency of oscillation and compare it with the calculated value.
2 Vary the power supply between 0 V and the maximum allowed and hence draw a graph of output frequency against power supply voltage. What important conclusions can you draw from the graph?

Soldering

In practical work in electronics, it cannot be emphasised too strongly that neat and logical arrangements and reliable solder joints are of paramount

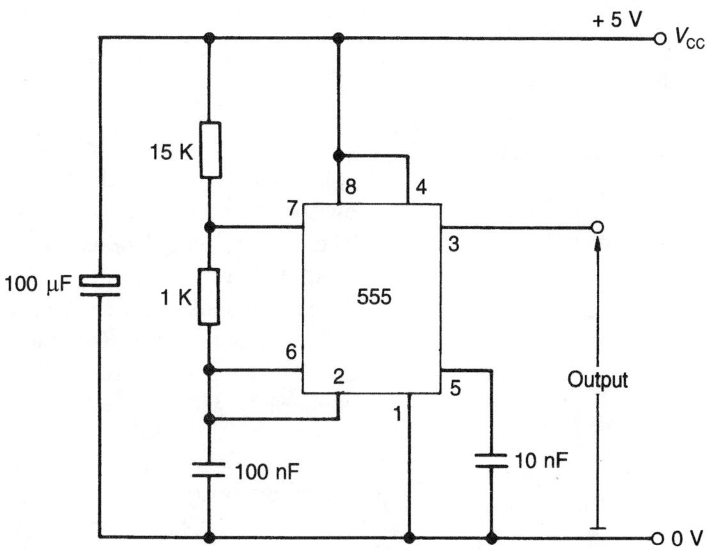

Figure 19.9 A typical 555 timer circuit

importance. Basically, there are four methods of making electrical connections: nuts and bolts, wire-wrapping, crimping and soldering. The latter is undoubtedly the most important.

Solder is an alloy of tin and lead which melts at around 180°C. When copper is heated, a layer of oxide forms on its surface which would normally prevent the solder adhering. To prevent this, a material called FLUX is applied which cleans the surface and prevents oxidation. Although flux can be applied separately, most solder used in electronics already contains flux inside it. This is commonly called 'multi-core' solder.

The following points should be noted when soldering:

1 Select the correct power rating of iron. 40 W is a good standard size; for pcb (printed circuit board) work, an iron of around 25 W would normally be used. For other work, note that too small an iron can lead to inefficient joints.
2 Ensure that the bit is not worn; pitted copper bits can be filed when cool until they are flat. Iron bits should not be filed.
  A dirty bit can be cleaned using a damp sponge. Most soldering iron holders have a well for the sponge in the base. Once cleaned the bit should be 'tinned' by allowing solder to run over it.
3 The solder must flow evenly over and between the surfaces being soldered. If the surfaces are properly tinned, clean and heated to the correct temperature, this should not be a problem.
4 When making the actual joint, the iron is *not* used to heat the solder; the iron heats the wire or component and this then heats the solder. Allow the solder to flow and give it time to cool properly to avoid dry joints.

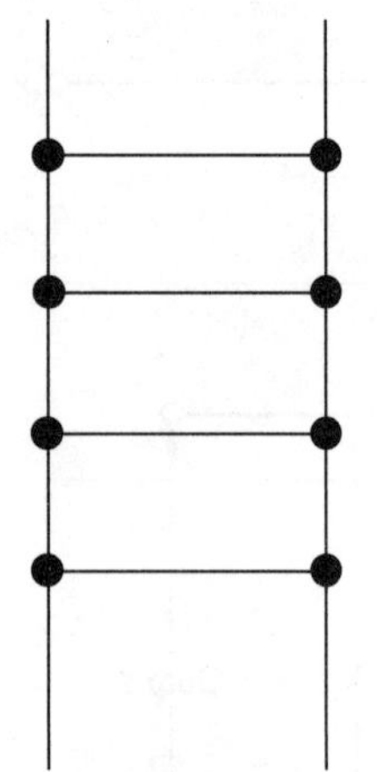
Figure 19.10 Simple soldering exercise

REMEMBER – LET THE SOLDER FLOW

5 Do not flick solder about; remember that solder melts at around 180°C – nearly twice the boiling point of water. It can damage your eyes permanently. Do not try to catch a soldering iron if it falls.
6 Mains irons should be earthed. Ideally, low voltage irons should be used (12 V or 24 V). All connections should be checked before use.
7 Some textbooks advise you to use a heat sink when soldering delicate components like transistors. In actual fact, this is usually unnecessary – if you're careful. Modern components are designed to cope easily with the heat applied during soldering.

Soldering exercises

Tools and equipment required

Soldering iron; holder with damp sponge; solder; solder board; radio pliers; side cutters; about 0.5 m of tinned copper wire (tcw).

Using tcw, make a ladder with four 'rungs' similar to that shown in Figure 19.10. The rungs should be about 2 cm wide and 1 cm apart.

Repeat the exercise using PVC covered wire. Try to waste as little wire as possible.

Using tcw, construct a pyramid as shown in Figure 19.11. The overall height of the pyramid should be about 8 cm. Try making a grid as shown in Figure 19.12.

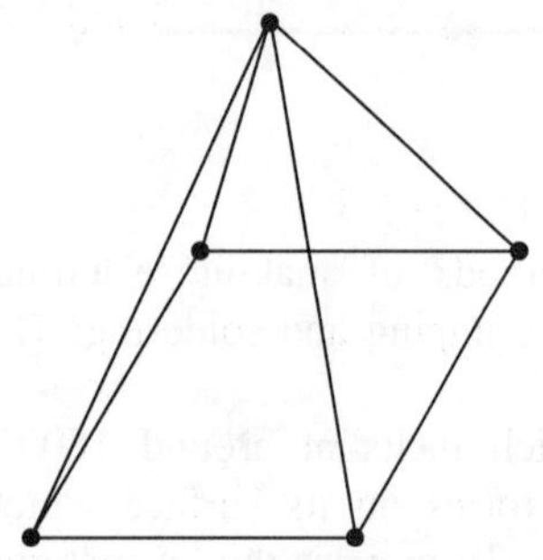
Figure 19.11

Using stripboard ('veroboard')

Almost without exception, it is best to design your circuit with the copper strips running *horizontally*.

1 Nominate a positive (V_{cc}) line and a negative line. Usually, these will be the top and bottom strips of copper on the panel.
2 Decide where the transistors and/or ICs go first.
3 Avoid placing components diagonally – it makes it much more difficult to trace faults or take readings and looks untidy.
4 Some connections can be made across the board using tcw – they are called 'jumpers'. This is acceptable but should be kept to a minimum.
5 Place the stripboard with the strips going away from you (rather than from left to right); you are then less likely to get solder in the channels between the strips. Press lightly with the iron on the copper for good thermal contact.
6 Let the copper strip melt the solder (rather than the iron directly); you then know you have the correct temperature. Put in sufficient solder, don't dab at the joint – put the iron on and keep it there – LET THE SOLDER FLOW!

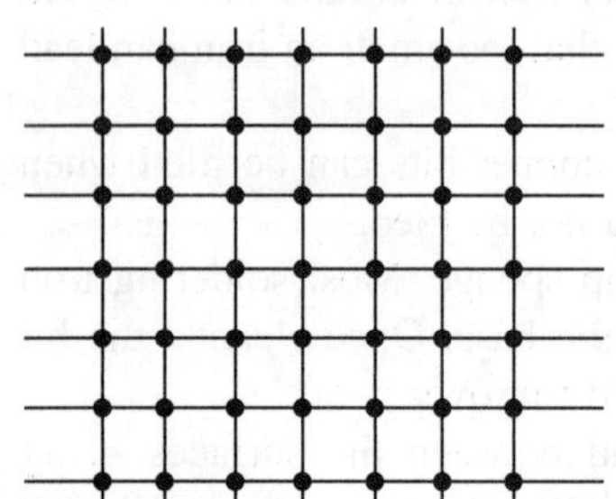
Figure 19.12

7 Some holes may have to be made (not all the way through!) in order to break the copper strip at various places. A special tool called a spot face cutter (or strip cutter) is available for this purpose. It's rather like a small drill with a handle on it.
8 ABOVE ALL! REMEMBER that the components are *not* mounted on the copper side. The components are mounted on the non-copper side and soldered to the copper tracks underneath.

Desoldering

Most components can be removed from a veroboard layout or pcb quite easily, simply by heating the solder and withdrawing the component, either by hand or with the aid of radio pliers. In some cases, it is useful to add more solder in order to help the old solder to flow; this is because the solder contains flux.

> Great care must be exercised when desoldering and it is wise to wear a pair of goggles. Some solder melts at around 180°C so getting it in your eyes could lead to permanent loss of vision.

For larger areas of solder, or when removing multiple-electrode devices (e.g. transistors or integrated circuits) it is useful to use a desoldering tool, or to use a solder sucker in conjunction with a standard soldering iron. It is possible to obtain a combined iron/desoldering tool but it is more common to find a solder sucker.

For integrated circuits, iron tips, the shape of the IC pin-out, are available. In the absence of such a device, 8-pin DIL devices can be removed as follows:

1 Use a solder sucker to remove the solder from each joint in turn. Where the solder persists, heat that joint while gently easing the device from the panel.
2 Clean the print up using the soldering iron, after the device has been removed.

Larger 14- or 16-pin DIL devices may be more difficult to remove as described above, in which case the use of a desoldering gun or solder sucker is strongly recommended. A desoldering braid is also available to 'soak up' the solder. It is rather like the copper braid which forms the outer of certain types of coaxial cable (the feeder from the aerial to a television set).

Components

Before looking at the following assignments, note the component details given in Figure 19.13.

Colour coding of connecting wires

It is suggested that connecting wires are colour coded as follows:

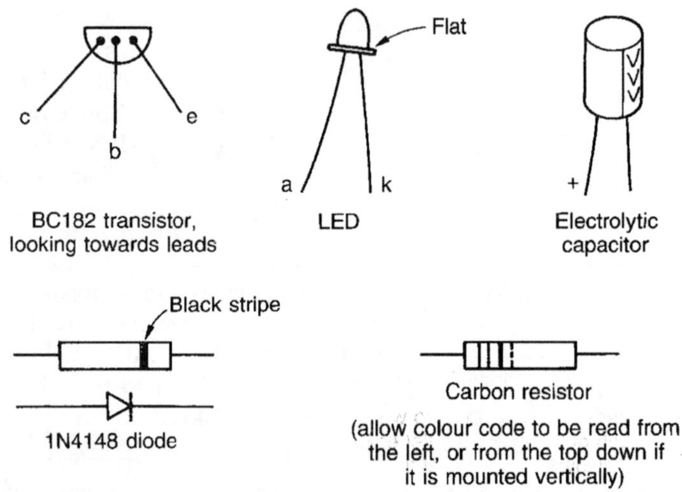

Figure 19.13 Some component details

RED – exclusively for positive supply lines
BLACK – exclusively for zero volt (0 V) lines
BLUE – for negative supply lines
GREEN – for earth lines

WHITE – for input signal lines
YELLOW – for output signal lines

As a rough guide, 15 cm of wire is sufficient for most connecting leads.

Signal generator system project

This simple project allows the constructor to use three different methods of implementing electronic circuits. A UJT oscillator is described and this can be built on stripboard. A sine wave generator is built up on a printed circuit board (pcb) prepared by the etch resist method while a Schmitt trigger uses a pcb using the photo resist method. The completed system may also be used in order to demonstrate some elementary fault-finding techniques as described in the next chapter.

A UJT ramp (or sawtooth) generator

The circuit uses a 2N2643 or 2N2646 unijunction transistor (UJT). The circuit diagram is shown in Figure 19.14, while Figure 19.15 shows a suggested stripboard layout. An alternative design may be used but inexperienced constructors would be well advised to follow the design shown.

The circuit may be checked by connecting the power leads to a suitable power supply unit (PSU). The actual voltage is unimportant as long as 15 V is not exceeded, and a 6 V supply has proved to be satisfactory. If an

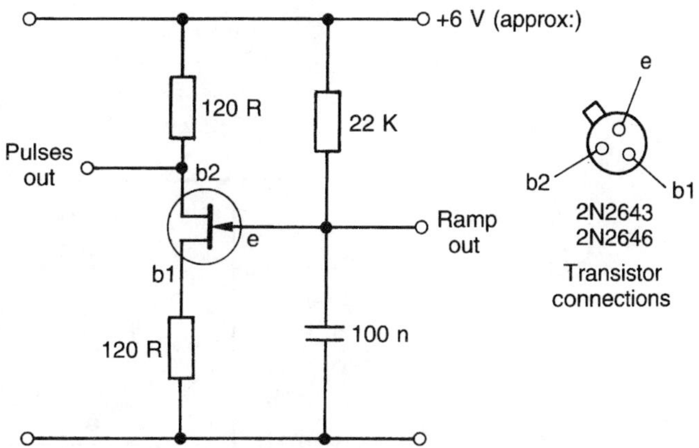

Figure 19.14 Circuit diagram of UJT ramp oscillator

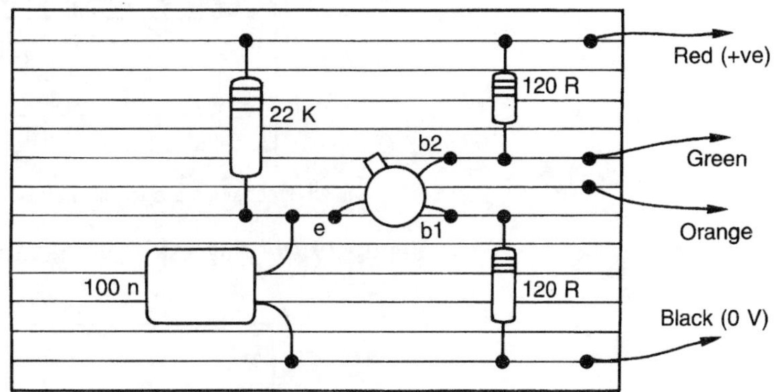

Figure 19.15 Suggested layout diagram for UJT oscillator

oscilloscope is now connected to the two outputs, base 2 should produce pulses and the emitter should produce a ramp waveform.

The sine wave oscillator

This circuit is a very simplified version of a phase shift oscillator. Figure 19.16 shows a pattern for the printed circuit board which should be produced using the etch resist technique.

The pcb design is not an ideal one; it has its roots in an old City and Guilds 224 assignment which, once completed, was consigned to the rubbish bin. Since this seemed a bit of a waste it was decided to try and use the pcb for something and so the circuit was designed around the pcb rather than the other way round. Accordingly, the merits of producing this simple pcb have been retained and, additionally, some use is also made of it. As an alternative, Figure 19.17 shows an improved pcb layout while

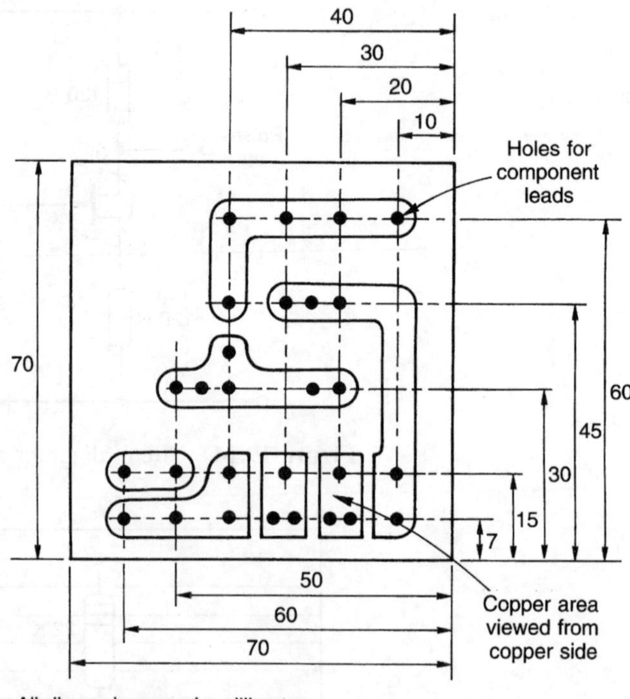

Figure 19.16 Suggested pcb pattern for phase shift oscillator

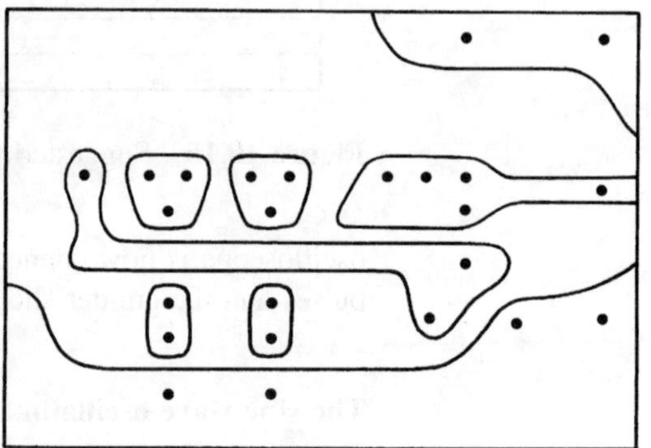

Figure 19.17 Alternative design for phase shift oscillator pcb

the component side is shown in Figure 19.18. Provision is made on this pcb for the addition of a variable frequency control.

1 Clean the copper side of the board thoroughly using wire wool or an ordinary pencil eraser.
2 Mark out the pattern in pencil making all measurements from one chosen corner of the panel. The panel may be supplied ready-cut; it

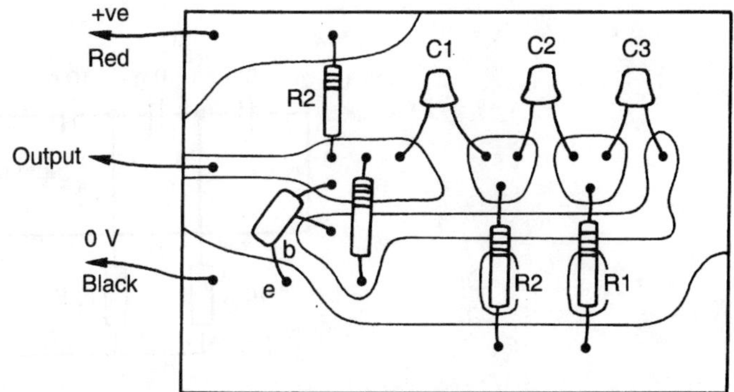

Figure 19.18 Component side of pcb

does not have to be exactly 7 cm square as long as all measurements are made from one chosen point.

3 Using an etch resist pen, fill in all the copper areas that are to remain after the etching process.

4 Remove the unwanted copper by placing the board in the etching solution.

5 Remove the etch resist material from the pcb with acetone or wire wool.

6 Drill out the holes and clean the panel.

7 The panel may now be populated as shown in Figure 19.19, the component values being as follows:

R1, R2 = 10 kΩ R3 = 4.7 kΩ R4 = 470 kΩ

C1, C2 and C3 = 10 nF (0.01 μF) Tr1 = BC182

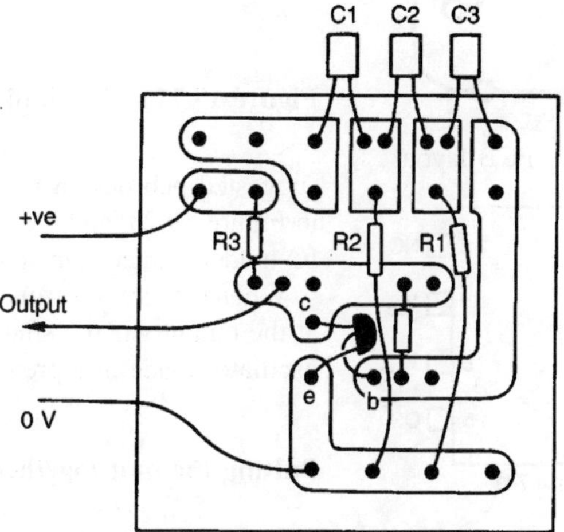

Figure 19.19 Component layout

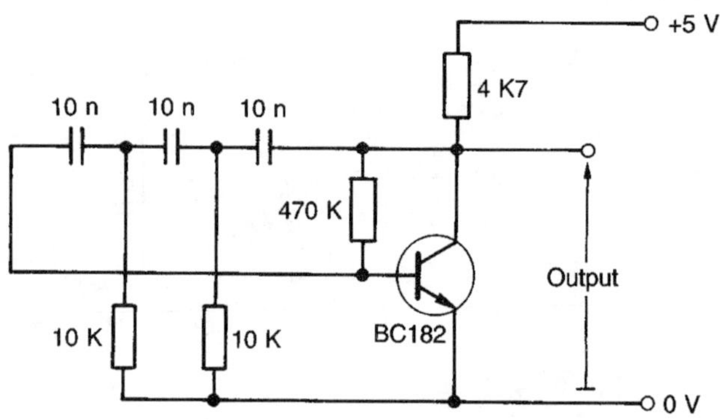

Figure 19.20 Circuit diagram of phase shift oscillator

The circuit diagram is shown in Figure 19.20.

The Schmitt trigger

The circuit should be implemented using the photo resist method of pcb construction. The circuit diagram is shown in Figure 19.21 and a suggested pcb design is given in Figure 19.22. The 741 pin-out is given in Figure 19.23, and it's a good idea to mount the device on an 8-pin DIL IC holder, rather than solder it directly onto the pcb.

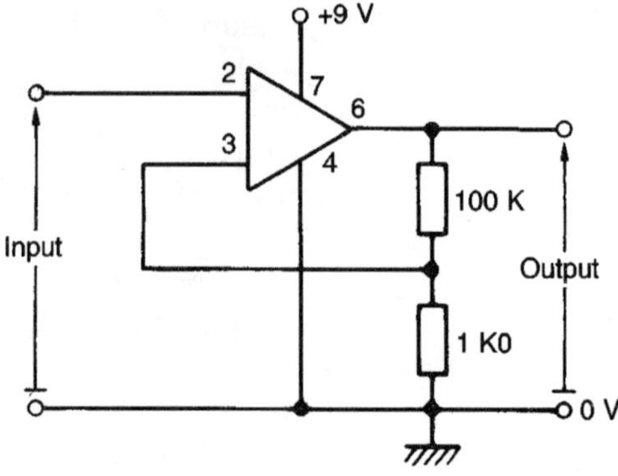

Figure 19.21 Circuit diagram of Schmitt trigger

Figure 19.22 PCB layout

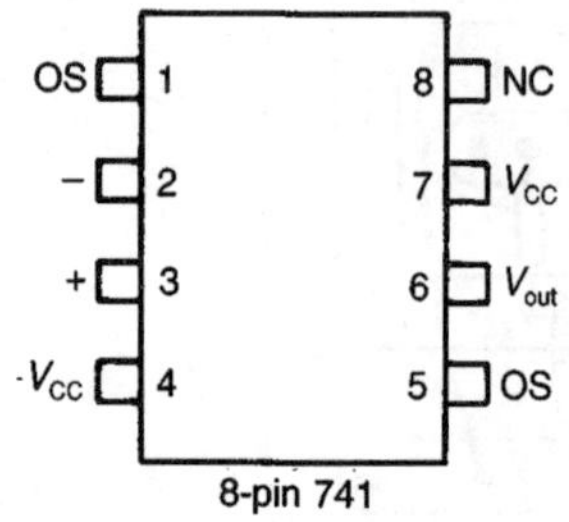

Figure 19.23 Pin-out of 741 IC

When the circuit is functioning correctly, a square wave will be produced at the output when a sine wave is connected to the input. The phase shift oscillator made in a previous assignment can be used to test this circuit.

Putting the unit together

The following notes will enable the ramp generator and sine wave oscillator to be modified so that their output frequencies can be varied.

Remember, however, that these are very basic circuits so that it is not possible to obtain variations over a large frequency range.

The UJT oscillator is the easiest to modify. With the component values given, the minimum resistance for the emitter which will enable the circuit to oscillate is about 12 kΩ. So, remove the 22 kΩ resistor, replace it with a 12 kΩ resistor and put a 100 kΩ potentiometer in series with it, as shown in Figure 19.24. This should give a frequency range between about 70 Hz and 700 kHz.

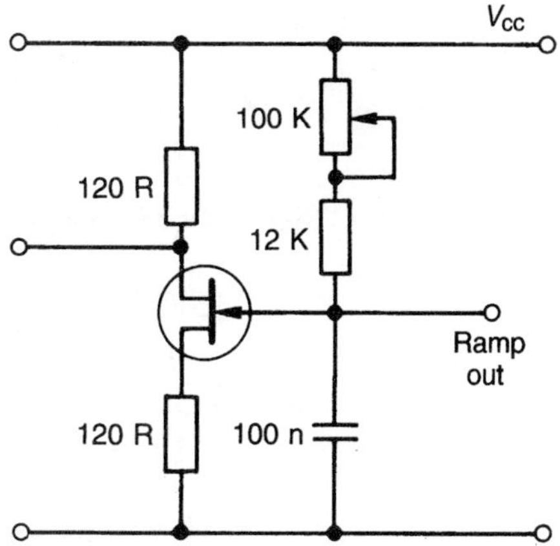

Figure 19.24 The modified ramp oscillator circuit

The phase shift oscillator cannot easily be made variable, but if a dual potentiometer is used, some frequency control can be achieved, though there may be some distortion present.

Begin by removing *one* of the 10 kΩ resistors in the feedback circuit and insert a 10 kΩ potentiometer. Vary this while the circuit is connected to the CRO and find, by experiment, the lowest resistance that will still allow the circuit to oscillate. This can be done by setting the potentiometer to the required position, removing it from the circuit and then measuring its resistance on the ohms range of a multimeter. Use the next preferred value resistor above and put this in series with the potentiometer. For example, it may be found that with 10 nF capacitors, 2.2 kΩ series resistors with a 10 kΩ dual potentiometer will allow some frequency adjustment. This arrangement is shown in Figure 19.25.

Metalwork

All three panels may be fitted onto a piece of aluminium sheet measuring 20 cm by 16 cm, bent at right angles along its length and down the middle. Holes are cut for power input and signal output sockets and for the frequency adjusting controls. This is shown in Figure 19.26.

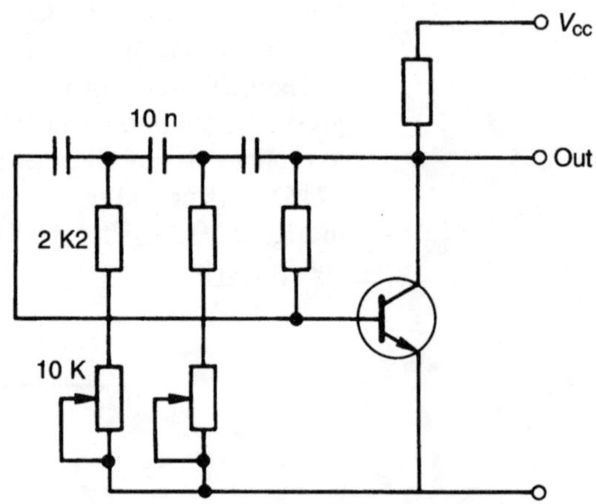

Figure 19.25 The modified phase shift oscillator

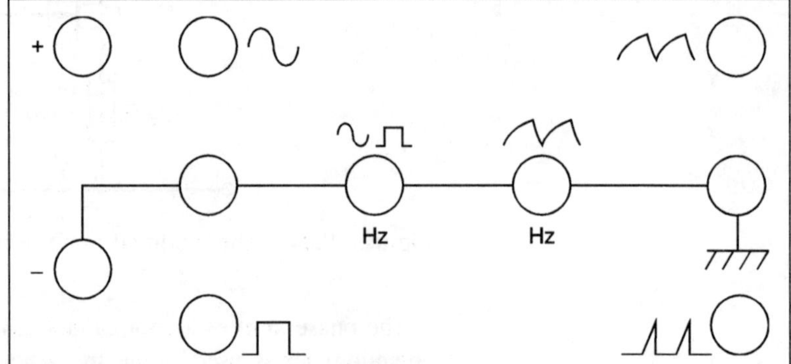

Figure 19.26 Suggested layout for front panel of signal generator

Sockets and controls

The power input sockets should be terminal posts with 4 mm sockets, red for the positive input and black for the negative (or zero volts) input. The other six sockets may be simple, flush mounting 4 mm sockets, black for the common outlets and any other colour but red for the others.

The frequency adjusting controls will normally be standard 10.5 mm fixings, but this should be checked before drilling and/or reaming. Suitable holes will need to be drilled in each of the circuit panels. Stand-off posts, nylon pcb mounts or just nuts and bolts may be used. All wires should be neatly routed and secured using spot ties, cable wrap or lacing, as appropriate.

This simple project introduces students to three different types of circuit construction (stripboard, etch resist and photo resist pcbs). The device produces sine, square and ramp waveforms as well as a pulse output. The completed unit is, therefore, a simple but useful source of four different types of test signal which may be used in the servicing and repair of other electronic equipment.

In addition, as it stands, or with modifications as desired, the unit may be used in the elementary study of simple fault-finding techniques. This is particularly appropriate in the light of the publication of the 2348/9-05 NVQ in Servicing Electronic Systems (Workshop), which requires some basic fault-finding ability. The next chapter deals with this subject in some detail.

20 Electronic fault finding

General fault-finding techniques

Fault finding is one of the most difficult techniques to acquire in electronics. It requires the application of theoretical knowledge, experience and practical expertise. Fortunately, there are many aids both theoretical and practical, and to conclude this book some basics are examined. First of all, there are several rules to bear in mind when learning fault-finding techniques, and these should be remembered at all times:

1 Always use service manuals and other relevant documentation; make notes.
2 Always adopt a logical approach.
3 Check – and then recheck.
4 The practice doesn't always fit the theory.
5 Things aren't always what they seem.
6 You can put a fault on a piece of equipment while attempting to repair it.
7 A second fault can occur on a piece of equipment, while looking for the first.
8 Finally: NEVER TAKE ANYTHING FOR GRANTED!

These are important rules you should think carefully about; unfortunately, only after some practice and experience will you take them seriously. You are encouraged to do so right from the start. It's no good buying a brand new computer and trying to make sense of it by trial and error; then, after hours of frustration, opening the instruction book. Have everything on your side right from the start. So, having introduced a few basic rules, let's look at some introductory fault-finding techniques.

A Schmitt trigger

Figure 20.1 shows the circuit of a Schmitt trigger introduced in the last chapter. The circuit contains only three components: a 741 op-amp and just two resistors. What could be easier?

The Schmitt trigger is an electronic circuit which switches between its output maximum and minimum at a critical level of input. The effect is, that, regardless of the input, the output is a square wave (assuming that the input varies above and below the critical levels).

The easiest way to check the circuit, therefore, is to apply a sine wave and use a CRO to monitor both input and output. Assuming there to be a fault present, we need to develop a strategy for locating it in the shortest possible time. In the case of this simple circuit, it wouldn't take long to replace all the components and hence cure the fault, but this is not good practice. It may work for this circuit, but supposing it had fifty ICs and hundreds of associated components, the method would clearly be untenable.

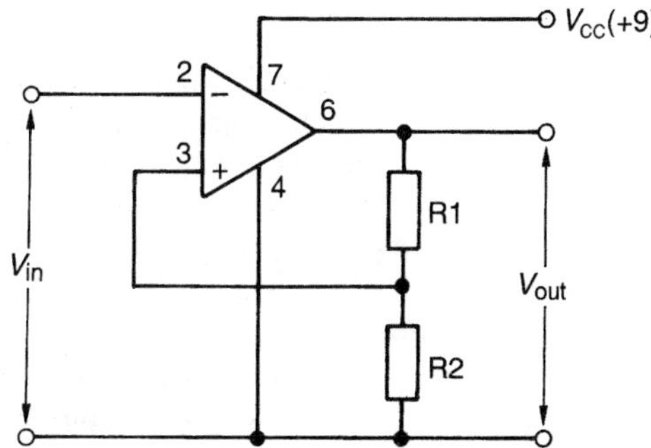

Figure 20.1 Simple Schmitt trigger circuit using a 741 op-amp

Here is one suggestion for tackling the problem.

Step 1

Let's assume that a power supply of 8 V is being used. Check that the correct voltages are present; typical voltages are given in Table 20.1. A good starting point is to measure across pins 4 (0 V) and 7 (V_{cc}). If all is well, move on to (step 2). If not, check other points for V_{cc}, gradually moving back to the power source. Note that it is important to measure across the pins of the IC and then work backwards; obtaining the correct voltage at source tells you only that it's present there, it doesn't mean it's on the IC pins themselves. There could be a dry joint at one of several places, the IC socket (where fitted) could be faulty, there may be a break in the print. Working backwards as suggested will reveal any such possibilities; further checks on the 'ohms' range (continuity) will reveal its location.

Table 20.1 Typical 741 voltages

Pin	*AC*	*DC*
2	2.2 V	4.24 V
3	0.01 V	0.02 V
4	0 V	0 V
6	2.48 V	2.78 V
7	0 V	8.0 V

Step 2

If 8 V is present across pins 4 and 7, check for a signal on pin 2 of the IC. Comments given in the last paragraph apply here; a signal on the input to the panel does not prove the presence of a signal on the chip itself. If it's absent, work backwards to locate it.

Step 3

If a signal is present on pin 2 and there is still no output at pin 6, it's almost certain that the IC is faulty. This should be checked by substitution.

Step 4

If there is still no output, it will be necessary to check elsewhere in the circuit. The possibility that either R1 or R2 is short circuit is about a million to one against (I've never known a carbon resistor go short circuit (S/C). Resistors tend to go high resistance or open circuit (O/C) and neither is likely in this case. Nevertheless, if all else fails, the resistor values should be checked.

The procedure is given in some detail here, in order to demonstrate a logical way of going about things. Generally, however, experience will eventually show you that you will need only to check for appropriate supply voltages and input signal and if they're all right then change the IC. The 741 is called an 'active' component, and they're the ones that usually fail.

The other two circuits in the signal generator system (phase shift and UJT oscillators) will be examined later on in this chapter.

Component faults

Having jumped in at the deep end with an example, let's take stock.

Resistors Carbon types may increase in value or go open circuit. Examples of both possibilities usually occur only when the component passes consistently higher current than it is designed for (unlikely – the manufacturers would use a higher power type) or momentarily higher current owing to a fault elsewhere. In signal circuits, where little current flows, resistors tend to outlive the equipment they form part of. Wirewound resistors frequently fail because of the heat they dissipate. It is almost unknown for any type of resistor to go S/C.

Capacitors All types can go S/C, O/C or low in value. Electrolytic types can leak (literally) and the physical manifestation of this is easy to see.

Diodes and transistors Signal types seem to go on forever; power types frequently go O/C or S/C.

Testing components

The easiest and quickest method of testing discrete (separate) components is to use a multimeter. There are two main types of meter, analogue and digital. In order to describe the procedure, we will use an 'AVO' meter (analogue) and a 'BECKMAN' meter (digital). A simple representation of an AVO is shown in Figure 20.2.

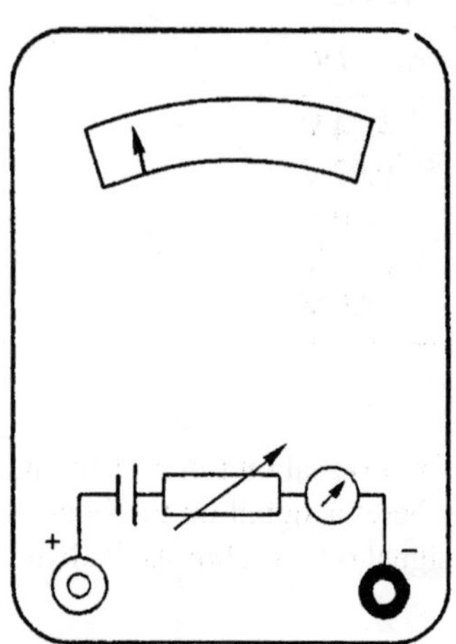

Figure 20.2 The AVO terminals showing how the battery is connected internally

The AVO has a 15 V internal battery as shown. Rather confusingly, the positive terminal of the meter is connected to the negative terminal of the battery. If a S/C is placed across the meter terminals (on the lowest ohms range), the needle of the meter should be deflected across the scale and read ZERO ohms on the right hand side. A good fuse is a virtual S/C, so this is one way of testing it. No movement of the meter needle indicates an O/C or a reading of infinity (∞) ohms.

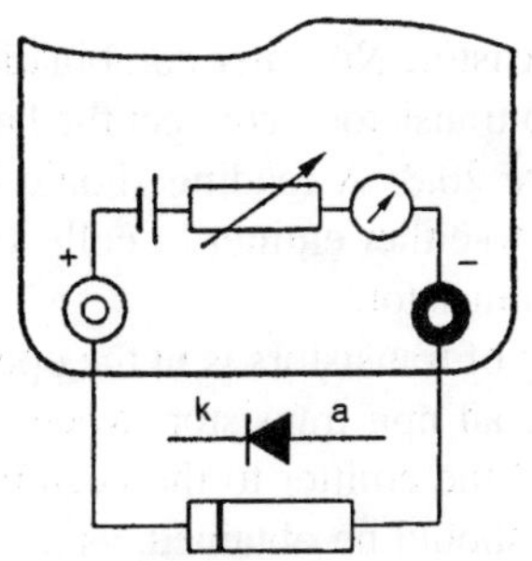

Figure 20.3 Testing diodes

Testing diodes

Set the AVO to the ohms $\times$ 1 range and connect the diode as shown in Figure 20.3.

With the cathode on the positive terminal of the meter, a reading of about 20 to 30 Ω, up to a few hundred ohms, should be obtained, depending on the type of diode. Connected the opposite way, the resistance should be several million ohms (MΩ) and it is usual to observe no reading at all.

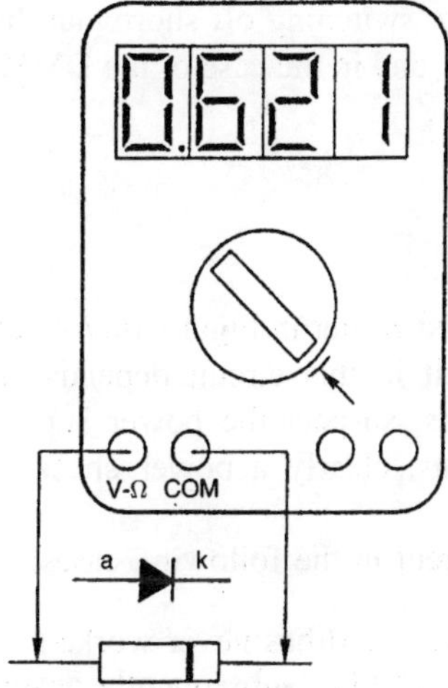

Figure 20.4 Testing diodes with a DMM

Use of a digital multimeter (DMM)

The procedure is similar to that described above, except that on *most* digital meters, the polarity of the internal battery is the same as the terminals on the front of it. Some DMMs (e.g. the Beckman) have a special diode testing circuit. Connected as shown in Figure 20.4 on the diode range (marked with the diode symbol) a reading of about 0.62 V should be obtained. In the reverse direction, the meter displays OL (overload).

With a standard (red) LED connected as shown in Figure 20.5, the meter should indicate a value of about 1.6 V. This may be slightly different for diodes of different types or different colours. In the reverse direction, the meter gives the OL signal. The diode itself should glow slightly during this operation.

Germanium point contact diodes such as the OA91 will give a reading of around 0.55 V. This may vary slightly from one diode to another. The value also varies as a function of temperature. A simple representation is given in Figure 20.6. Bridge rectifiers and many other types of diode may also be measured in this way.

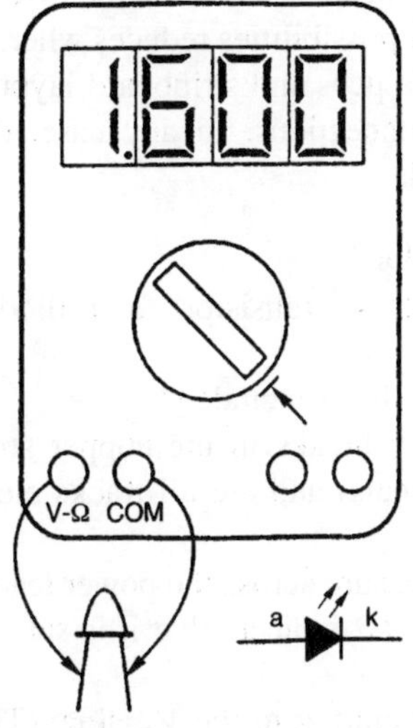

Figure 20.5 Testing LEDs with a DMM

Transistor testing

Testing transistors is more difficult to describe than the procedure for testing diodes, because there are so many different types. The special unijunction transistor (UJT) is discussed later when we look at a ramp oscillator circuit which uses this device. Here, we will just consider bipolar, collector–base–emitter type transistors. Common npn transistors include the BC107, BC108, BC109 and the familiar BC182, BC183 and BC237A plastic encapsulated types. The pnp equivalents are BC186, BC187, BC478 and BC212. To test npn transistors, connect as shown in Figure 20.7 using an AVO type meter.

The base of the transistor is connected to the negative of the meter. When the positive lead is connected to either emitter or collector, a reading

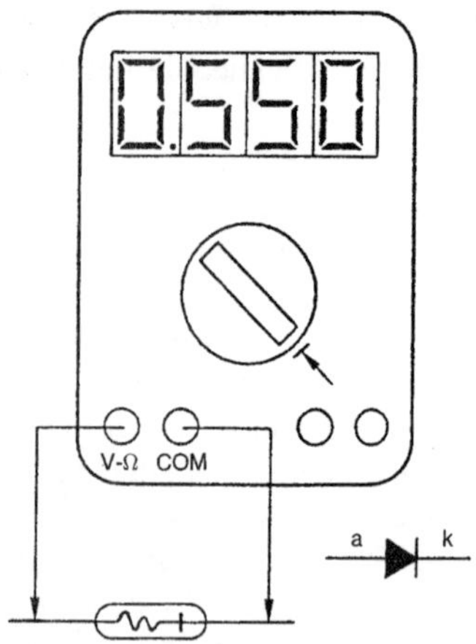

Figure 20.6 Testing an OA91 signal diode

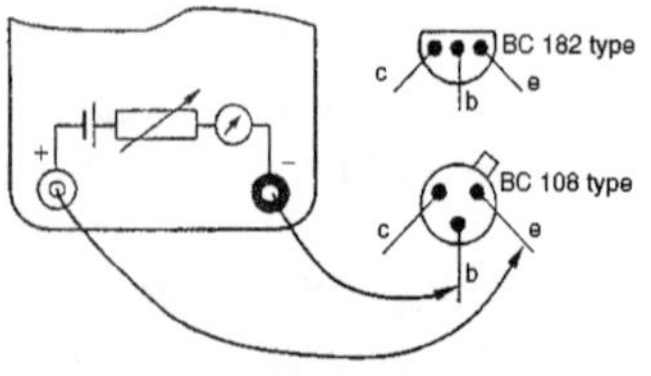

Figure 20.7 Testing npn transistors with an AVO

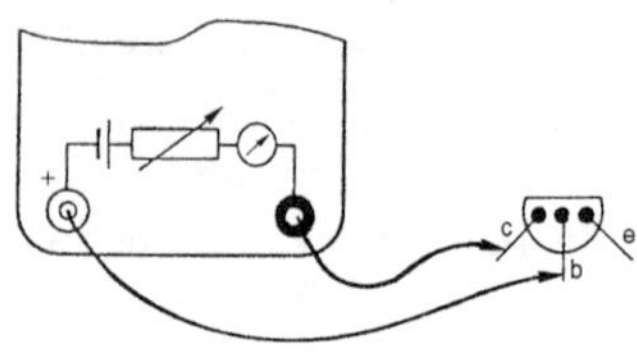

Figure 20.8 Testing pnp transistors with an AVO

of between 20 and 30 Ω indicates a good transistor. No other combination of connections should give a reading. For pnp transistors, connect the base to the positive lead of the meter – see Figure 20.8. A reading should be obtained when the negative lead is connected to either emitter or collector. Any other readings would indicate a faulty transistor.

When using a digital multimeter, the testing of transistors is in the opposite sense to that described for the AVO. For an npn transistor, when the base is connected to the positive terminal and the emitter to the negative, a reading of about 0.62 V on the diode range should be obtained, as in the case of the analogous diode tests.

Meters should always be returned to the OFF position when not in use or being transported. In the case of the AVO, switching off shorts out the meter movement protecting it from vibration, and in the case of the DMM, it saves the battery.

Testing a simple digital circuit

We now return to the astable circuit introduced earlier in Figure 19.7 – see page 237. The procedure for finding a fault in this circuit depends on what the symptom is. If neither LED lights, suspect the power supply connections; if one or other (or both) lights, clearly a power check is unnecessary.

The fault-finding approach is rather different in the following cases:

1 Where a circuit has been built from scratch and has never worked.
2 Where a circuit has been known to work and has subsequently developed a fault.

In the first case, components could have been inserted in the wrong place or the wrong way round. The number of possibilities reduces where a manufactured pcb has been used. Home-made pcbs and stripboard layouts are obviously more susceptible to faulty connections. In any case, the following preliminary checks should be made:

1 Are all the components of the correct value?
2 Are (electrolytic and tantalum) capacitors, transistors and diodes connected the right way round?
3 Have all necessary wire links ('jumpers') been included?
4 On stripboard designs, have all necessary breaks in the copper strip been made (e.g. between capacitor terminals) and are all tracks electrically insulated?
   If all visual checks appear satisfactory, measure across the power leads. On an AVO meter, readings above a few kΩ indicate that it is safe to connect up the circuit.
5 Check that 5 V exist from each transistor emitter to the V_{cc} line. (The emitter may seem to be connected to 0 V but this is the only way to be sure.)

6 Short each transistor collector down to 0 V in turn. In each case, the LED should glow. If it doesn't, recheck LED polarity, resistor value and then all associated connections.
7 If all is well, one LED should be 'ON' and the other 'OFF'. The one that is off should come on when its base is connected to V_{cc} *via a 1k resistor.* If either transistor fails this test, check it with a meter as described earlier.
8 If each transistor switches as it should, check the capacitor polarity, and finally check these components by substitution.

In most cases, the above checks will reveal the fault.

The phase shift oscillator

Now let's look at a typical fault-finding approach on a phase shift oscillator, similar to that described in the last chapter. Figure 20.9 shows a slightly different circuit for such an oscillator, but the original could be used just as well. If that's the case, typical measurements can be generated from a circuit, known to be functioning correctly.

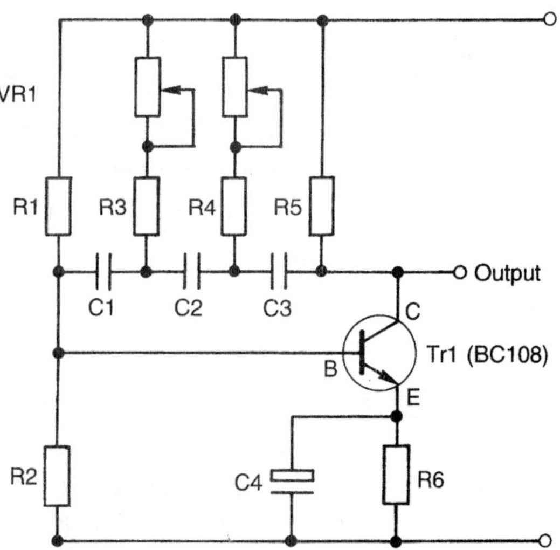

Figure 20.9 A phase shift oscillator

With the values of components shown, an 8 V power supply produces a reasonably undistorted output; typical voltage and other measurements are shown in Table 20.2.

The difference in readings obtained by the use of the two types of meter are negligible as we would expect. The slight differences do exist, of course, but no experienced engineer would read much into them. Note that the difference between base and voltage readings produces a value of about 0.55 V to 0.6 V, most silicon transistors have this base/emitter voltage when they are working correctly.

Table 20.2 Typical measurements for a phase shift oscillator

Meter	*Supply volts*	V_c	V_b	V_e	*Current drawn*	*Frequency*
AVO	8 V	4.0 V	0.80 V	0.25 V	1.50 mA	365 Hz
DVM	8 V	4.28 V	0.88 V	0.26 V	1.55 mA	365 Hz

The fault-finding procedure is similar to that described for the Schmitt trigger at the beginning of this chapter.

Step 1

Assuming a power supply of 8 V, first check the transistor voltages. Using a DVM, the collector voltage should be about 4.3 V. A reading of 0 V would indicate a power supply problem, a break in the print or a dry joint on the transistor. A very low voltage would seem to indicate that the transistor is S/C. Zero volts would suggest that R5 is O/C. It is also possible that R2 is O/C so that the transistor is fully ON dropping the full supply voltage across R5.

Step 2

If the collector voltage is at 8 V, no current is being drawn and the transistor is either O/C or there is no voltage on the base. If the base voltage is normal then it looks as though the transistor is faulty. If the base voltage is zero, suspect R1 to be O/C or look for a break in the print.

Step 3

Check the emitter voltage. If it is zero then either C4 is S/C or the transistor is faulty.

Step 4

If the transistor voltages appear normal, but there is still no output, check the voltages at the junctions of C1/C2 and C2/C3. These should be almost at the supply potential (8 V). Anomalous readings here would suggest a problem with R3 or R4, or the potentiometer and associated wiring.

Generally, as we have seen before, if there is a fault, it is most likely to be the transistor itself.

The UJT ramp oscillator

This oscillator is based on a unijunction transistor (UJT). This device can be checked with a DVM as shown in Figure 20.10.

With the meter on the diode range and the positive lead on the emitter, a good UJT produces about 0.8 V, base 2 to emitter and about 0.9 V, base 1 to emitter. There should be no reading at all in the reverse direction. With the positive lead on b1 and the negative lead on b2, a reading of about 1 V (nothing in the reverse direction) indicates a good UJT.

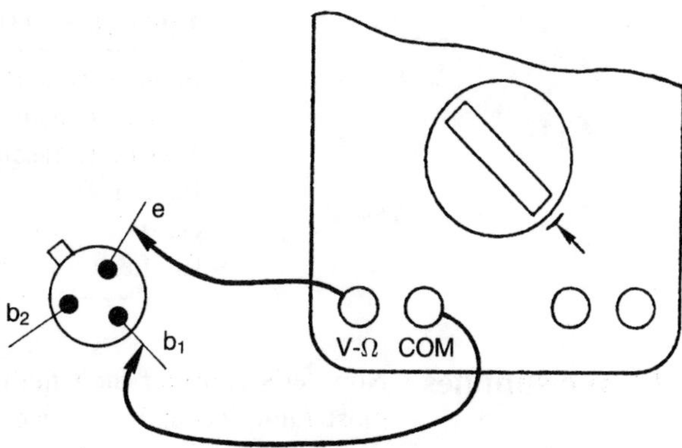

Figure 20.10 Testing a unijunction transistor (UJT)

Using an analogue meter (20 kΩ/V)

On the ohms × 1 range and the meter's negative lead on the emitter, a reading of about 30 Ω to each base should be expected, and no reading at all using the opposite polarity.

With the negative lead on base 2 and the positive lead on base 1, a reading of about 50 Ω should be obtained, nothing in reverse.

The UJT ramp generator

A simple circuit was shown in the last chapter (Figure 19.14). A similar circuit, which has switchable frequency ranges, is shown in Figure 20.11.

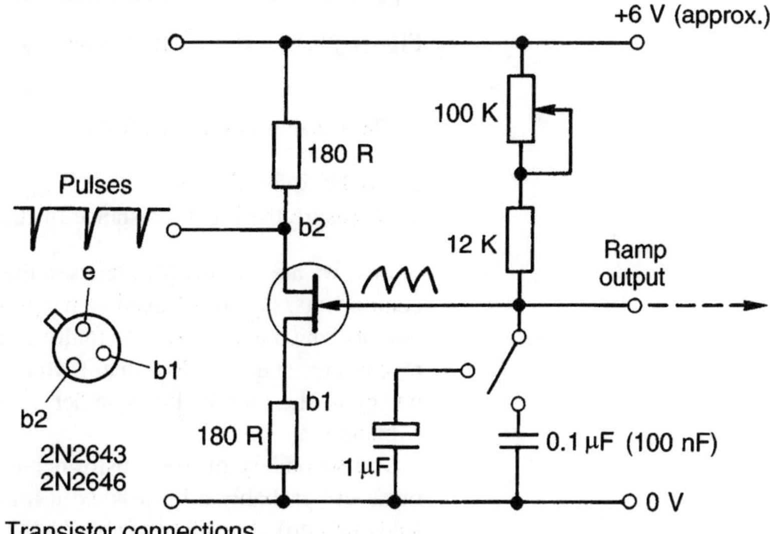

Figure 20.11 A UJT ramp generator

With an 8 V supply, the readings in Table 20.3 will be obtained when the circuit is working normally.

Table 20.3 UJT oscillator measurements

Input resistance	5.2 kΩ
Current consumption	1.5 mA to 2.3 mA
Maximum frequency	667 Hz
Base 1	0.41 V
Base 2	7.65 V
Emitter	4.23 V

Power supplies

Now let's consider fault finding in power supplies. It is true to say that most faults occur here since all the power to the equipment must pass through its power supply; this is where most heat is generated and heat is the biggest enemy of electronics. Figure 20.12 shows a simple basic power supply circuit.

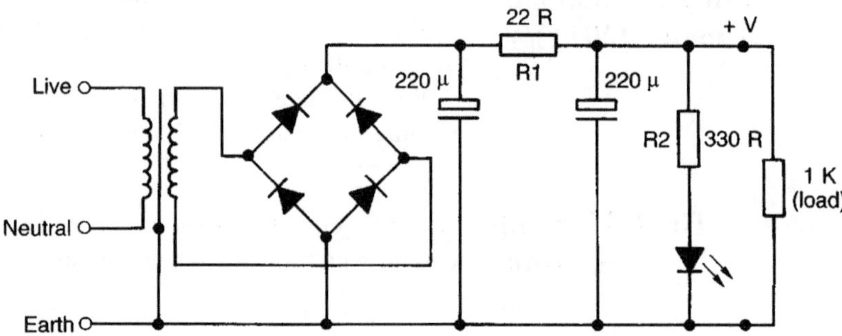

Figure 20.12 Basic simple power supply circuit

The starting point is as follows:

1 Make a visual check.
2 Measure the input resistance of the circuit.

In 1, we are looking for breaks in the panel or print, missing or damaged components, or wires frayed or hanging off. In 2 we are looking for short circuits which may betray the faulty component(s) directly, or which would reveal a resistance so low that it would be unwise or dangerous to connect power to the circuit. Items under 1 are fairly obvious, 2 requires some explanation.

The secondary of the transformer will measure only a few ohms at the most and probably a lot less, so it must first be disconnected (one of the leads will do). A reading of about 8 kΩ or so (in this circuit) across the AC input to the bridge rectifier would indicate a normal reading and the unit safe to power up. A reading of less than 6 kΩ should make you suspicious, and around 3 kΩ suggest a short circuit diode or smoothing capacitor. A reading of a few hundred ohms in such a circuit should set the alarm bells ringing – you would not even consider powering the circuit up.

If a short circuit is suspected (because of an abnormally low reading), it is a simple matter to check all four diodes (as described earlier), and the smoothing capacitors on a resistance range. Short circuit components usually reveal themselves in circuit, but, because of parallel paths, the only way to be certain is to remove the component (again, by disconnecting one end only) and test it out of circuit.

If all is well, the circuit can be powered up and checks made roughly according to Table 20.4.

Table 20.4 Suggested fault-finding procedure for simple power supply

	Test	*Result and action*
	(Visual check)	NOF (No obvious fault)
	(Check input resistance)	8 kΩ – safe to switch on
1a	Does LED light?	YES – see later NO – suspect LED or R2
2	Measure voltage across LED	0 V – suspect S/C LED or O/C R2
		+V – LED is O/C
3	Measure across load	Normal – check LED and/or 330 Ω on meter
		0 V – fault further back in circuit
4	Check at 'hot' end of 22R	+V – R1 is O/C
		0 V – fault further back in circuit
5	Check transformer secondary	OK – suspect primary and/or plug fuse
6	Check primary	OK – suspect fuse or wiring
1b	Does LED light?	YES – assume low voltage
2b	Measure across load	Low volts – suspect O/C diode or cap
3b	Put CRO across load	Measure ripple frequency:
		100 Hz – check smoothing capacitors
		50 Hz – one or more diodes O/C

The half split method

This method involves the isolation of part of a circuit (not necessarily half) in order to determine in which part the fault may lie. To take a simple example, suppose a power supply contains several smoothing capacitors in parallel as shown in Figure 20.13.

If a measurement at X – Y indicates a S/C, then it could be the fault of any of the components. Making a break at point D verifies the integrity of half the capacitors immediately (unless two or more are S/C which, fortunately, is very rare). The half split process is then continued until the

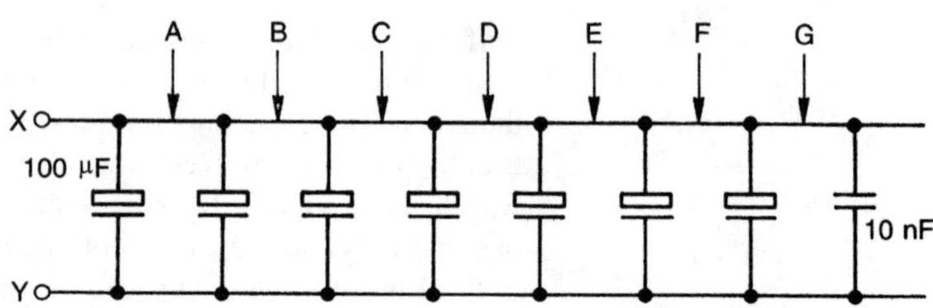

Figure 20.13 The half split method of fault finding

faulty component is located. In reality, this technique may be confounded by any of the following:

1 It may not be practicable to isolate parts of the circuit in order to make the measurements, although some electronic equipment is made up from units ('panels') which simply plug in, making the process much easier. Cutting printed circuit tracks is not usually a good idea.
2 All the components may not be equally reliable.
3 All checks may not take a similar amount of time.

Common sense and some experience will help you to decide when the method is appropriate, and when it is not.

Audio amplifiers

A typical, small signal amplifier is shown in Figure 20.14.

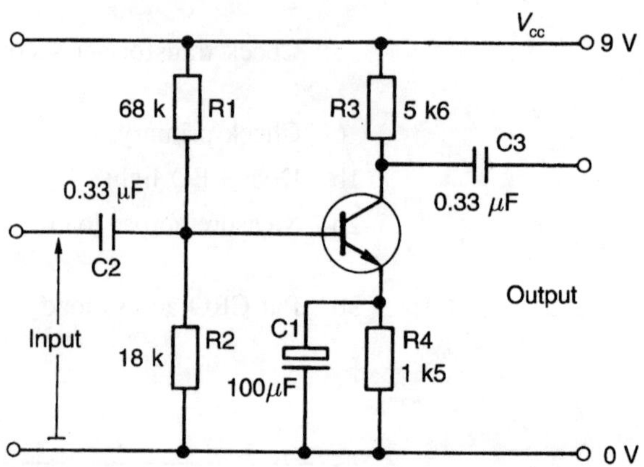

Figure 20.14 A typical, small signal audio amplifier

With a power supply of 9 V, typical voltages are as shown in Table 20.5.

Possible faults

1 R1 O/C. If R1 is O/C, there will be no base voltage so the transistor is switched off, there is no collector current, so the collector measures V_{cc}. So: $V_b = 0\,V$, $V_c = 9\,V$ and $V_e = 0\,V$.

Table 20.5 Transistor voltages

	AVO	*DVM*
V_b	1.7 V	1.92 V
V_e	1.2 V	1.29 V
V_c	4.2 V	4.27 V

2 R2 O/C. In this case, the full V_{cc} is applied to the base via R1; the transistor conducts very hard so that collector/emitter current is high; there is a large voltage drop across R3, V_c is low, and with a greater voltage drop across R4, V_e increases.

3 R3 O/C. V_c *must* read 0 V (except that if an AVO is used, some current will flow through the meter). Base current still flows and emitter voltage follows the base voltage less the 0.6 V drop across the base/emitter junction. Typical voltages (using an AVO) are: $V_b = 0.7\,\text{V}$, $V_e = 0.1\,\text{V}$, $V_c = 0.1\,\text{V}$.

4 R4 O/C. With R4 O/C, the base/emitter circuit is O/C so there is no base current and the transistor is switched off. There is no emitter/collector current so no voltage is dropped across R3 and $V_c = V_{cc}$. Typical voltage readings are: $V_b = 1.85\,\text{V}$, $V_e = 1.35\,\text{V}$ (because AVO meter is connected) and $V_c = 9\,\text{V}$.

5 C1 S/C. With C1 S/C, R4 is effectively S/C as well, base current increases and there is a larger volts drop across R3 so V_c decreases. There is also a larger volts drop across R1 since I_b increases. Typical voltage readings are: $V_b = 0.6\,\text{V}$, $V_e = 0\,\text{V}$, $V_c = 0.25\,\text{V}$.

6 Transistor S/C collector/base. With the collector/base junction S/C, additional current flows through R3/R2 so there is a larger volts drop across R3 reducing V_c; there is also a larger volts drop across R4 so V_e increases. Similarly, there is a larger volts drop across R2 so V_b also increases. Typical voltages are: $V_b = 2.4\,\text{V}$, $V_e = 1.8\,\text{V}$, $V_c = 2.4\,\text{V}$.

7 Transistor S/C base/emitter. With this fault, current flows through R4 and R1 via the S/C junction so I_b increases causing a larger drop across R1 and a corresponding decrease in base voltage. There will be no collector current, so that $V_c = V_{cc}$. This fault is easy to determine since V_b and V_e are equal. Typical voltages: V_b and $V_e = 0.1\,\text{V}$ and $V_c = 9\,\text{V}$.

8 Transistor O/C base/emitter. There will be no base current and no forward bias, lower volts drop across R1 so V_b increases. With no base current there is no collector/emitter current, no volts drop across R3 so collector volts at V_{cc}, and emitter at 0 V. Typical readings therefore: $V_b = 1.85\,\text{V}$, $V_e = 0\,\text{V}$ and $V_c = 9\,\text{V}$.

9 C1 O/C. With the emitter capacitor O/C, almost 100% negative feedback occurs and the amplifier has a much reduced gain. There will be little or no changes in the DC voltages. Typical voltage readings: $V_b = 1.7\,\text{V}$, $V_e = 1.2\,\text{V}$, $V_c = 4.2\,\text{V}$. The best way to detect an O/C emitter capacitor is to monitor the input and output of the amplifier

and bridge the suspect component with a known, good capacitor. If the output increases dramatically, then the original capacitor is obviously at fault.

10 C2 or C3 O/C. The DC bias conditions will not be affected, but either no signal appears on the base or, even with a signal on the collector, there is still no output from the amplifier.

Gain and bandwidth

To complete this section, typical readings (Table 20.6) of gain and bandwidth are given for this amplifier.

Table 20.6 Typical readings for the amplifier

Input p-p	*Output p-p*	*Gain @ 1 kHz*	*0.707 of O/P*	f_{low}	f_{high}	*Bandwidth*
20 mV	3.2 V	160	2.26 V	115 Hz	330 kHz	329.885 kHz

Logic circuits

Let's start by considering an AND gate as shown in Figure 20.15 with its truth table.

Suppose that input A is permanently at logic 1, because of an internal fault. This would be described as being 'stuck at one' or 'S-A-1'. The truth table would then be as shown in Table 20.7 below.

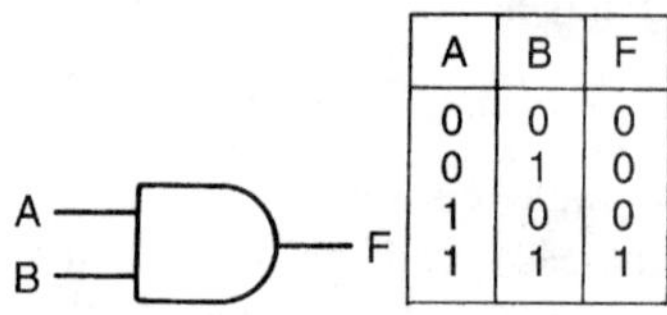

A	B	F
0	0	0
0	1	0
1	0	0
1	1	1

Figure 20.15 An AND gate with its truth table

Table 20.7 Truth table for an AND gate as it would appear if input A were S-A-1

A	B	F
0	0	0
0	1	1
1	0	0
1	1	1

Applying a zero to A and a logic 1 to B will reveal this fault. If the output is S-A-1, then the output column, (F), would contain all 1s regardless of the inputs applied. Most faults in logic circuits are revealed by going through the relevant truth table and identifying anomalous readings.

Note that this fault-finding section is presented as a guide only; true ability in the subject comes almost entirely from PRACTICE.

The NVQ in Servicing Electronic Systems Level 2 requires more in the way of adopting the correct *procedures* rather than expecting students at this level to have achieved experienced skill in actual fault finding.

Answers

Chapter 1

Multiple choice questions

1	B	2	A	3	B	4	D	5	D
6	A	7	D	8	B	9	A	10	D
11	D	12	B	13	A	14	D	15	D
16	B	17	B	18	A	19	D	20	D
21	C	22	A	23	D	24	C	25	A

Chapter 2

Questions

1 1111 Ω 2 9.375 Ω 3 40 Ω
4 50.11 Ω 5 100 Ω 6 0.49 A
7 1.58 A 8 6 V 9 0.5 A
10 Across 5 Ω = 6.25 V, across 3 Ω = 3.75 V 11 1 V
12 40 V 13 20 V 14 Electron
15 (a) copper (b) silicon (c) brass
(d) PVC (e) gold
(f) polystyrene, mica, paper, air (g) aluminium
(h) mercury
16 8.1 Ω 17 5×10^{-7} Ω m.
18 1.3 m 19 2.5×10^{-8} m^2
20 First calculate R_t from $V/I = 13.33\ \Omega$
Calculate R for nichrome from $R = \rho l/A = 20\ \Omega$
Calculate R for manganin from $\dfrac{1}{R_m} = \dfrac{1}{13.33} - \dfrac{1}{20}$.
Then: $R_m = 40\ \Omega$
Next, calculate area, A, for manganin:

$$A = \rho l/R = 1.26 \times 10^{-7}\ \text{m}^2$$

$$A = \pi r^2 \quad \text{therefore: } r = \sqrt{A/\pi}$$

$$\text{so:} \quad r = 2 \times 10^{-4}\ \text{m}$$

$$\text{and:} \quad d = 4 \times 10^{-4}\ \text{m} = 0.4\ \text{mm}$$

Multiple choice questions

1	B	2	A	3	C	4	B	5	C
6	B	7	C	8	B	9	A	10	C

Chapter 3 Questions

1 $C = \dfrac{8.85 \times 10^{-12} \times 2.5}{0.1 \times 10^{-3}} = 0.22\,\mu F$
2 44 pF
3 0.133 mm
4 1.77 nF
5 45 pF
6 Electrolytic
7 (a) polarity; voltage working; ripple current; ensure absence of AC
(b) polycarbonate and tantalum
(c) paper
8 8.3 μF
9 3.3×10^{-5} μF
10 Works out to 320 V (approx.) so would probably choose one of about 400 V working
11 (a) $Q = 2.5 \times 10^{-6}$ C
(b) $Q = 4.7 \times 10^{-5}$ C
So (b), the ceramic capacitor holds more charge
12 50 μF
13 6 μF
14 7.3 μF
15 0.047 μF
16 (a) 100 nF = 0.1 μF (b) 100 pF = 0.1 nF
(c) 10 000 pF = 0.01 μF (d) 1000 μF = 0.001 F.
17 3330
18 (a) $Q = 4.8 \times 10^{-4}$ C
(b) 6 μF, $V = 80$ V; 4 μF, $V = 120$ V.
(c) 6 μF, 1.2×10^{-3} C; 4 μF, 8.0×10^{-4} C
(d) 200 V
19 ($C = 180$ pF); $Q = 1.8 \times 10^{-8}$ C
20 10 μF, 2 mC; 20 μF, 4 mC; 70 μF, 14 mC

Multiple choice questions

1 B	2 D	3 B	4 B	5 A
6 B	7 B	8 C	9 B	10 A
11 C	12 A	13 C	14 A	15 D
16 A	17 B	18 B	19 C	20 B

Chapter 4 Multiple choice questions

1 D	2 A	3 D	4 A	5 D
6 C	7 B	8 D	9 C	10 A
11 A	12 B	13 A	14 C	15 B

16 D	17 B	18 A	19 C	20 A
21 C	22 B	23 B	24 D	25 C
26 D	27 C	28 B	29 C	30 A

Chapter 5 Questions

1 Conduction, convection and radiation
2 (a) A heat sink is used to conduct the heat away
(b) A metal heat sink is used to conduct the heat away, is large in area to set up convection currents and painted black to maximise radiation
(c) A fan is used
(d) A large surface area is produced by using fins
3 (a) Very little change in resistance
(b) Quite a large change in resistance
(c) Resistance reduces with an increase in temperature causing more current to flow, causing a further increase in temperature – thermal runaway sets in which would damage the transistor if left unchecked
(d) PTC – increases resistance as temperature increases
NTC – decreases resistance as temperature increases
4 (a) kg (b) newton (c) joule
5 (a) 100 N (b) 1500 J
6 Power = 5 W
7 (a) Kg (b) N (c) joule (d) joule
(e) joule (f) N (g) joule
8 $R_t = 25.08\ \Omega$
9 (a) 0.2 s (b) 2.5 m/s
10 (a) 5×10^{-6} m (5 μm)
11 2000 m
12 330 m/s
13 0.01 m; 5 waves.
14 0.48 m.
15 6.9×10^{14} Hz.
16 30°
17 (a) red (b) violet
18 (a) White, yellow, cyan, green, magenta, red, blue, black
(b) Red, green and blue
(c) White = red + green + blue
Yellow = red + green
Cyan = green + blue
Magenta = blue + red
19 Black
20 Because different coloured light is refracted through different angles in a glass prism

Multiple choice questions

1 C	2 A	3 C	4 C
5 B	6 B	7 B	8 A
9 A	10 B	11 A	12 C

Chapter 6 Questions

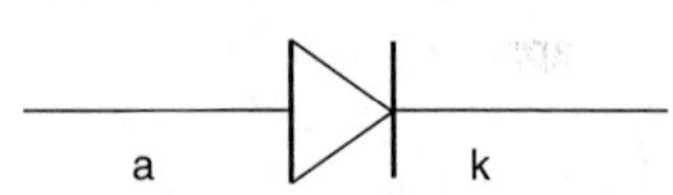

Figure A1 BS symbol for a diode showing anode (a) and cathode (k)

1 See Figure A.1
2 See Figure 6.2
3 Any value between about 470 μF and 4700 μF is typical
4 See Figure 6.6
5 Regulator or stabiliser
6 Any value between about 30 kHz to 150 kHz
7 Must be greater than audio frequencies (>20 kHz) so it cannot be heard
8 Mixed with the audio signal for recording and used as an erase signal
9 (a) $3\frac{1}{4}$ i.p.s. (b) $1\frac{5}{8}$ i.p.s.
10 Because it's more effective at slower speeds

Chapter 7 Multiple choice questions

1 D	2 D	3 C	4 B	5 D
6 B	7 B	8 C	9 A	10 A

Chapter 8 Questions

1 (a) The output from the DC amplifier is varied either by the set speed control or the amount of the feedback obtained from the DC generator. In each case, it's the *difference* between the two inputs of the differential amplifier that is proportional to its output and which drives the DC amplifier.
(b) The speed may be automatically maintained by using the feedback signal from the DC generator.
(c) A differential amplifier is one which amplifies the *difference* between its two inputs.
(d) A potentiometer with each end connected to +V and 0V via resistors of appropriate value, the slider being connected to one of the differential amplifier inputs.

2 (a) The phase discriminator has an output which is proportional to the difference in phase between its two inputs.
(b) VCO = Voltage Controlled Oscillator; the control voltage originates in the phase discriminator and reaches the VCO via a filter and DC amplifier.

(c) The filter is a low pass one which removes the high frequency components of the phase discriminator output.
(d) The buffer prevents loading of the VCO either by the phase discriminator or any load to which the output of the system may be connected.

3 (a) The trace on the screen is produced by the bombardment of the electrons on the inside of the tube. The spot so produced is made to move by the action of a ramp generator, changing the spot into a horizontal line, the length of which is determined by the output of the X amplifier. If the frequency of the timebase is high enough, the afterglow of the phosphor and the effect of persistence of vision, makes the trace appear stationary.
(b) The process which causes electrons to leave a heated cathode is called thermionic emission.
(c) After the trace appears on the CRO screen the rapid return of the electron beam to its starting point is called the flyback.
(d) The CRO trace remains stationary because a sync. pulse derived from the input waveform is applied to the CRO timebase to keep it in step.

4 (a) The line frequency is 15625 Hz and is derived from there being 25 frames, each of 625 lines, every second on the TV screen. $625 \times 25 = 15625$.
(b) TV picture flicker was reduced by the use of interlaced scanning.
(c) The coils around the tube neck which produce the horizontal and vertical scanning are called 'scan coils'.
(d) The inter-carrier amplifier extracts the audio information from the video IF signal.

Multiple choice questions

1 D	2 A	3 D	4 B	5 A
6 A	7 C	8 D	9 C	10 B
11 C	12 A	13 C	14 A	15 B

Chapter 9 Multiple choice questions

1 C	2 D	3 C	4 A	5 C
6 B	7 C	8 C	9 A	10 D

Chapter 10 Multiple choice questions

1 C	2 B	3 A	4 A	5 C
6 B	7 C	8 C	9 B	10 A
11 C	12 B	13 D	14 A	15 B

Chapter 11 Multiple choice questions

1 B	2 B	3 A	4 C	5 C
6 D	7 D	8 D	9 A	10 D

Chapter 12 Multiple choice questions

1 D	2 B	3 A	4 B	5 D
6 A	7 A	8 C	9 D	10 A
11 B	12 D			

Chapter 13 Questions

1 (a) 1912 (b) 1876 (c) 34092 (d) 105.222

2 (a) 50: 10% = 5; 15% = 7.5; 30% = 15, 75% = 37.5
(b) 88: 10% = 8.8, 15% = 13.2, 30% = 26.4
75% = 66

3 (a) (i) 3288 → 3300 (ii) 3945 → 3900
(iii) 1672 → 1700 (iv) 630 → 630
(v) 29.5 → 30 (vi) 0.0881 → 0.088
(b) (i) 3.14159 → 3.14 (ii) 0.815 → 0.82
(iii) 15.673 → 15.67 (iv) 8.999 → 9.00
(v) 0.0763 → 0.076 (vi) 144.2 → 144.20

4 (i) $6492 \rightarrow 6.492 \times 10^3$ (ii) $1100\,000 \rightarrow 1.1 \times 10^6$
(iii) $560\,000 \rightarrow 5.6 \times 10^5$ (iv) $22 \rightarrow 2.2 \times 10^1$
(v) $0.055 \rightarrow 5.5 \times 10^{-2}$ (vi) $0.0075 \times 7.5 \times 10^{-3}$
(vii) $27\,000\,000 \rightarrow 2.7 \times 10^7$
(viii) $59\,637 \rightarrow 5.9637 \times 10^4$ (ix) $1.33 \rightarrow 1.33 \times 10^0$
(x) $0.00009 \rightarrow 9 \times 10^{-5}$

Note Although the answer to (ix) is correct, normally it makes sense just to write down the number, in this case 1.33.

5 (a) (i) 1 kΩ ± 10% → 900 Ω to 1100 Ω
(ii) 27 kΩ → 24 300 Ω to 29 700 Ω
(iii) 56 kΩ → 6.16 kΩ to 50.4 kΩ
(iv) 220 kΩ → 198 kΩ to 242 kΩ
(v) 3.3 MΩ → 2.97 MΩ to 3.63 MΩ
(b) (i) 910 kΩ ± 5% → 864.5 kΩ to 955.5 kΩ
(ii) 75 Ω → 71.25 Ω to 78.75 Ω
(iii) 22 kΩ → 20.9 kΩ to 23.1 kΩ
(iv) 2.2 MΩ → 2.09 MΩ to 2.31 MΩ
(v) 15 kΩ → 14.25 kΩ to 15.75 kΩ

6 (i) $R = V/I$ (ii) $Q = CV$ (iii) $I = P/V$
(iv) $I = \sqrt{P/R}$ (v) $c = \sqrt{e/m}$

Chapter 14

Questions

1 (a) 22 kΩ (b) 4.7 kΩ (c) 3.3 MΩ (d) 150 Ω
(e) 680 kΩ (f) 27 Ω (g) 10 kΩ (h) 390 Ω
2 (a) brown, grey, orange
(b) grey, red, black
(c) blue, grey, brown
(d) orange, orange, red
(e) white, brown, yellow
(f) green, blue, green
(g) yellow, blue, brown
(h) brown, black, black
(i) green, blue, black
(j) red, violet, brown
(k) red, red, orange
(l) yellow, violet, yellow
3 (a) 82 Ω (b) 22 Ω (c) 470 Ω
4 Red, violet, black (27 Ω)
5 6.82 mA

Multiple choice questions

1 B 2 A 3 D 4 D 5 A

Chapter 15

Questions

1 The X-axis represents the independent variable
2 Gradient = y-axis value divided by x-axis value
3 An exponential
4 4 kHz
5 (a) 3333 Hz (b) 2 cm (c) 80 Hz
(d) 50 μs/cm (e) 67 Hz (f) 5 ms/cm

Multiple choice questions

1 A 2 C 3 D 4 D 5 C

Chapter 16

Questions

1 (a) Lead–acid
(b) Manganese dioxide

(c) 1.6 V
(d) Lead peroxide

2 (a) Zinc (b) Black
(c) Brown (d) Slate grey

3 Each positive plate is placed adjacent to two negative plates

4 (a) Water = 1.0 (just floats)
(b) Sulphuric acid = 1.28
(c) Turpentine < 1.00 (the float sinks)

5 See Table 16.2

6 (i) What it's made of (i.e. its resistivity)
(ii) its length (iii) its cross-sectional area
(iv) its temperature

7 (a) Electromagnetic effect – the magnetic field produced by a current-carrying conductor
(b) The chemical effect – electrolysis
(c) Heating effect
(d) Magnetic effect – a current-carrying conductor in a magnetic field experiences a force
(e) Heating effect
(f) Chemical effect – electrolysis
(g) Heating effect
(h) Magnetic effect
(i) Magnetic effect

Multiple choice questions

1 B 2 B

Chapter 17 Questions

1 (a) 3 (b) 5 (c) 12 (d) 15
(e) 195 (f) 103 (g) 239

2 (a) 1011 (b) 101101 (c) 10000
(d) 101 0011 (e) 1111 0101 (f) 100 1101
(g) 1 1000 1000 (h) 1100 1111 (i) 101010
(j) 1111 1001 (k) 1 1111 1111 (l) 11 1110 1000

3 (a) 1011 (b) 11100 (c) 100110 (d) 11010
(e) 10110 (f) 110110 (g) 1001011 (h) 111100
(i) 1100111 (j) 1111000

4 (a) 10 (b) 110 (c) 110 (d) 1010
(e) 10101 (f) 1010 (g) 111 (h) 100100
(i) 100 (j) 101111

Chapter 18 Multiple choice questions

1	D	2	A	3	C	4	D	5	C
6	A	7	C	8	B	9	A	10	A
11	A	12	B	13	B	14	C	15	A

Index